Eco Home

Eco Home

Smart ideas for sustainable New Zealand homes

MELINDA WILLIAMS

PENGUIN BOOKS

PENGUIN

UK | USA | Canada | Ireland | Australia
India | New Zealand | South Africa | China

Penguin is an imprint of the Penguin Random House group of companies, whose addresses can be found at global.penguinrandomhouse.com.

Penguin Random House New Zealand

First published by Penguin Random House New Zealand, 2018

3 5 7 9 10 8 6 4 2

Design by Cat Taylor © Penguin Random House New Zealand
Front cover photographs by Emma-Jane Hetherington (top and left)
and Simon Devitt (right)
Back cover photographs by David Straight (top) and Duncan Innes (bottom)
Prepress by Image Centre Group
Printed and bound in China by RR Donnelley Asia

A catalogue record for this book is available from the National Library of New Zealand.

ISBN 978-0-14-377153-1

penguin.co.nz

FSC www.fsc.org MIX Paper from responsible sources FSC® C144853

CONTENTS

CHAPTER 1

WHY MAKE AN ECO HOME?

Sunlight falls on the thermal mass of a concrete floor at a Titirangi home by Dorrington Atcheson Architects. Photo: Emma-Jane Hetherington

The eco house is the first one awake every morning. Even before dawn breaks, it gets to work for the people who live inside it. In the bedrooms and living spaces, blinds silently draw back to let the first sunbeams in, kick-starting the sleeping family's circadian rhythms so they wake naturally. As the sun rises, it warms the insulated eco-concrete slab floor in the living room — an expanse of thermal mass that absorbs heat that will be slowly released later in the day to keep the indoor temperature stable.

Upstairs, the family members stir as their rooms brighten. While Sam heads off for a solar-heated shower using a low-flow showerhead that uses half the water of a conventional one, Emma opens the double-glazed bifold doors, and steps out onto the roof deck into the morning air. As she takes a few deep breaths, a tiny copper skink that has crept out to warm itself on a paver skitters back into the shelter of a 'lizard hotel', one of several habitats for native creatures placed around the garden.

The children are in the kitchen for breakfast, boiling eggs on the energy-efficient induction cooktop. The built-in under-bench waste bins are separated for regular recycling, soft-plastics recycling, an odourless compost bin, and a small, almost empty bin for landfill. Later in the week, the compost bin will be emptied into the worm farm in the garden, and compost tea will be brought back to nourish the herb garden on the kitchen windowsill.

Although there are no heaters on and no windows open, it's the perfect temperature inside, and the air smells fresh and clean. That's thanks to the house being super-insulated and airtight, with an in-built ventilation system that continually exchanges air in the house. The walls are clad in naturally oiled wood that emits no toxic gases, bringing a warm softness to the interior. Leafy indoor plants help filter the air and give the interior a sense of connectedness to the outdoors.

As Sam heads to the garage, he throws a load of washing into the machine, which is fed from the house's rainwater collection tanks. A greywater system gives the used washing water a second life as it feeds out to the garden drip-feed system.

The family car is an electric vehicle, which has been charging overnight on the built-in fast-charge port in the garage. Sam's job requires travel during the day, so he takes the car. The kids join the walking bus to the local school as it passes, while Emma heads for the nearby train station. Today she's working in the city, but two days a week she works from home, in a study that converts to a guest bedroom when built-in wall-beds are folded down.

After the front door locks, the sensor lights switch off automatically. Throughout the day, a sequence of timers turn off power-hungry household electronics while activating appliances like the dishwasher and washing machine to make optimal use of the power generated by the house's photovoltaic roof tiles, and the 'solar paint' on the exterior walls. Through a smartphone app, family members can monitor the house's temperature, and remotely switch on the heat pump to warm the house before they return

Architects Justin and Louise Wright of Assembly Architects designed their super-insulated home to be warm, dry and healthy for their growing family. Photos: Simon Devitt

During the summer, the house opens up to let in natural light and air. In winter, its high insulation levels keep it cosy at low cost. Photos: Simon Devitt

home — in the unlikely event that it needs it.

In the afternoon, when the family returns, the LED lights in the home adjust their colour from the cool blue light of the morning to a warm yellow light, mimicking the changes in natural daylight. Dinner is made using fresh vegetables picked from the community permaculture garden shared by the neighbourhood, which includes beehives, a free-range chicken coop and even a frog pond. The family sits outside to eat, at a big table of recycled wood, as monarch butterflies cluster around a late-season bloom of proteas in the garden and tūī jostle at a nectar feeder.

Is this the perfect modern Kiwi eco home? No — but that's only because there's no such thing. If you're an idealist, this absence of a perfect model to strive towards might seem dispiriting at first. But the good news is there's no such thing as the perfect eco home because there are as many different ways of building a home designed to work in synchronicity with both its inhabitants and the natural world as there are people to live in them. The one described above is just one possibility of how a modern eco home might work, using technologies that already exist.

When you're building or renovating, dozens of factors will influence the decisions you make about your home, from the local climate and landscape, your budget, your sustainability priorities and your site, to how you want your home to look and feel, the size of your family, whether you use your home as an office or as a source of income, your maintenance tolerance, local bylaws and restrictions, and much more.

Our home is the biggest investment most of us ever make, and, increasingly, Kiwis want to make sustainable choices when we build or renovate. In the 2016 Colmar Brunton *Better Futures* report, 90 per cent of Kiwis said they have a medium or high commitment to living a sustainable lifestyle — and that percentage is growing year-on-year.

So what does it mean to build or transform a home into an 'eco home'?

The first thing an eco home must be is healthy for the people who live in it. That means warm, dry and well ventilated. Second, it should be built using materials that have minimal negative impact, in a way that produces as little waste as possible, and designed so that it requires few resources to run, thus minimising its contribution to climate change, pollution and environmental degradation. Ideally, it will be a building that is productive rather than consumptive — for example, generating more power than it uses. Third, it will be designed to make it easier for its inhabitants to live in a sustainable and productive way. And, finally, it should fulfill the promise of the word 'ecology', which refers to living relationships — to help the people who live in it feel more connected to the natural world they live in; a part of it, not apart from it.

Ultimately, creating a modern eco home is a complex juggling act that will likely require compromises and creative thinking. But the hard work is worth it, because it will bring the reward of a healthy and planet-friendly home that will be a great pleasure to live in for you and future residents for generations to come.

WARM, DRY, WELL VENTILATED

Building and living in an eco home benefits your health, comfort and pleasure. Warm, dry, mould-free homes have direct payoffs in fewer sick days, better sleep and greater comfort. Over time, the lower energy requirements save money, while eco-features have been shown to increase the resale value of a home.

New Zealand homes are notorious for being poorly insulated. Almost half our homes don't reach an evening average of 18 degrees Celsius, the minimum healthy temperature recommended by the World Health Organization. Our cold, damp, mouldy homes have sparked a chronic respiratory disease crisis in New Zealand, with thousands of children suffering from asthma, bronchiolitis, rheumatic fever and pneumonia every year. Older people suffer, too, with cold homes increasing the risk of circulatory disorders, stroke and heart trouble. Because we have put up with this state of affairs for so long, we often have trouble moving beyond the mindset that cold homes are normal, and that the answer is to throw an extra blanket on the bed, rather than expect more of the buildings we spend 70 per cent of our time in.

Part of the reason that New Zealand's homes are so poorly insulated is that until the 1960s all insulation had to be imported, so only the most expensive homes had it. It wasn't until 1978 that the government introduced minimum requirements around the insulation levels of new homes — levels that today are still poor by international standards. Although the government has spent nearly half a billion dollars in subsidies towards retrofitting homes with ceiling and floor insulation through the Warm Up New Zealand programme, up to half of existing homes still have little to no insulation at all.

On top of this, our existing housing stock is in serious need of upgrading. A study by the Building Research Association of New Zealand (BRANZ) showed that as many as 87 per cent of owner-occupied houses in this country need immediate repairs to prevent health and safety problems, and the numbers are likely to be as high or higher for rentals.

Adding pressure to the situation is the rising cost of houses, land and building, sparked by a severe housing shortage. Home ownership has declined steadily since the early 1990s, which is less to do with fewer people wanting to own their home, than with fewer people being able to afford to. Some respite may come in the form of the 2018 Labour-led coalition government KiwiBuild policy, which promises to deliver 100,000 'modern affordable homes' within 10 years.

When the first British immigrants arrived in New Zealand in the late 1700s, they were used to living cheek-to-jowl in terraced housing. To their eyes, the 1000 sqm parcels of land doled out by the government to build their chilly sheds on were, if not quite little slices of paradise, at least the first step on the road towards it. That traditional Kiwi 'quarter-acre dream' of a big house on a big suburban plot has endured ever since, but now, as our cities face intensive land pressure, our population ages, we have children later, our work habits change, and more people embrace apartment living and home-sharing, it's an increasingly irrelevant vision for the future.

The future that our young and yet-to-be born children will experience as young adults is likely to be radically different from the present. The nature of work is changing dramatically as automation and artificial intelligence eliminate many traditional jobs. A solidly built, future-focused sustainable home that has a measure of independence in terms of generating low-cost or free power, food and water may be a gift you can pass on to your children that allows them to live differently to how we do today — less at the mercy of corporate employment.

Modern, sustainable homes that are warmer, drier, healthier, more comfortable, more affordable and better suited to our lifestyles — the eco homes of the future — may be quite different to the ones we're used to living in. And that's without even talking about 'saving the planet' yet.

The Ira Street House in Wellington was designed to be an affordable, eco-friendly family home that makes optimal use of free natural resources. Photo: David Straight

MINIMUM IMPACT, MINIMUM WASTE, MINIMUM RESOURCE USE

Owning a home is the most tangible opportunity we have to make a powerful, practical statement about what we care about. In New Zealand, buildings are responsible for around 20 per cent of the country's energy-related greenhouse gas emissions, with homes making up the majority of that percentage. Buildings also consume 40 per cent of raw materials, and (not surprisingly) generate 40 per cent of our country's waste. By changing the way we build homes, we can make a big difference to our resource use, and our pollution and waste creation.

Over recent years, the real-world effects of climate change have become devastatingly clear. The previous four years have been the hottest on record, and we have already crossed the threshold of 400 (parts per million) of carbon dioxide in the atmosphere, the level at which scientists agree that the effects of the increased carbon dioxide will start to become significant.

In 2017, New Zealand was ravaged by extreme weather events in one of the wettest and hottest years on record. The 'Tasman Tempest' — the tail end of tropical cyclones Cook and Debbie — caused landslides and widespread flooding across the North Island in March and April. Many places received record or near-record rainfall days and months, while conversely Christchurch recorded its driest November on record and large parts of the western and southern North Island reached drought conditions. In their annual climate summary, the National Institute for Water and Atmospheric Research (NIWA) described it as 'a year of weather extremes'.

In the United States and the neighbouring regions, the double hit of Hurricane Harvey and Hurricane Irma caused billions of dollars of destruction across Texas, the Caribbean and Florida. These huge weather events are expected to increase as global temperatures rise, bringing more damage to our homes, properties and livestock, risks to our health through heat sickness, waterborne diseases, flooding-related deaths and injuries, and disruption of our connection with the land we live on.

In the same week Hurricane Irma hit, an international study reported in *The Guardian* showed microscopic plastic particles in 83 per cent of drinking water across five continents. The Great Pacific Garbage Patch of marine plastic debris (which has now grown so large that it's actually two patches — an eastern and a western one) is largely made up of these microscopic pellets of plastic and chemical sludge, suspended in the water. In 2017, two-thirds of the Great Barrier Reef suffered from coral bleaching — caused by rising sea temperatures — and some biologists say the damage is so extensive that this natural wonder may now never recover. Half of the world's natural forests have been cut down for agriculture and building materials. Over-grazing, pollution and monoculture is leading to severe soil degradation around the world. Over 100 species of wildlife become extinct every day in what the

Critical Ecosystem Partnership Fund (CEPF) calls a 'biodiversity crisis'.

Living in a country like New Zealand, which is relatively unpolluted by global standards, we can easily be misled into thinking that these challenges are less severe than they really are, or that they're somehow removed from us and beyond our ability to affect. Perhaps if our daily environment were as obviously degraded as the thick haze of air pollution in Beijing, or the endless roadside litter of India, our sense of urgency would be stronger.

Because when it comes to greenhouse gas emissions, despite the world's perception of New Zealand as a clean, green paradise, we don't have much to be proud of. While other Western countries have substantially reduced their net greenhouse gas emissions since the early 1990s, a report by former Parliamentary Commissioner for the Environment Jan Wright showed that New Zealand's net emissions actually *grew* by 64 per cent between 1990 and 2015. That makes us the second-highest greenhouse gas emitter per GDP unit in the OECD, according to the 2017 OECD Environmental Performance Review, and the fifth highest per capita. The 2017 Labour-led government has a target of New Zealand achieving net zero carbon emissions by 2050. If we're going to get there, everyone has to be in it together.

NEW ZEALAND HOMES BY THE NUMBERS

- 1.8 million houses
- 63 per cent owner-occupied
- Less than 15 per cent of people lived in cities when the first European settlers arrived in New Zealand; 85 per cent of people live in urban areas in New Zealand today (1.4 million in Auckland alone)
- 58.4 per cent growth in apartment living from 2006 to 2013
- By 2061, New Zealand will have 1.44 million aged people, with increased accessibility needs

FEELING A SENSE OF ECO-DESPAIR?

Sometimes the scale of the ecological challenges facing the world can be so overwhelming that it's hard to see how what we do as individuals makes a difference. And while that idea in itself is flawed — small actions do add up, and change at the governmental level is usually spurred by change at the community level — after watching yet another news report about global warming, even the most assiduous recycler can feel helpless.

To improve your resilience when it comes to your influence on the state of the world, imagine two circles, one inside the other. The smaller circle is your 'circle of influence'. That's the area where you can take direct action and make a tangible difference. For example, while there might be no direct action you can take to protect the Maui's dolphin (unless you're a West Coast fishing-trawler captain), you can preserve and encourage local biodiversity by building micro-habitats and planting native plants in your own backyard. (See Chapter 14: The Landscape for details.)

The larger circle is your 'circle of concern'. That's where you add your voice, attend protests, write letters, sign petitions, spread the word, and vote. That's important work, too. But to relieve that feeling that nothing you do matters much, focus on working within your circle of influence.

ENCOURAGING SUSTAINABLE LIVING

When we think of decreasing our carbon footprint, we often envision changing our shopping choices. And while shopping with reusable bags and buying natural cleaning products do have a positive effect, these are small actions in the big scheme of things. On the other hand, the way our homes are built and landscaped can dramatically change how much we consume in fossil fuels and natural resources like water, how much waste and pollution we create, and whether the way we live restores rather than degrades the environment.

To build sustainability into our homes brings the natural processes of the world into our daily lives in a way that makes the impact of our choices impossible to ignore, as we more readily can when we are disconnected from them.

When we have a composting system in our home, we are directly responsible for transforming our waste into something that can be returned helpfully to the land. When we are aware of how a cloudy day reduces our solar-power generation and bumps up our reliance on the grid, we no longer take electricity for granted in the way we do when it comes from some distant power station. When we use greywater systems, we are forced to consider what the substances we put down our plugholes might do to creatures in our own garden, or the plants we are growing to eat. All of a sudden, the direct impact of our choices becomes tangible.

The iconic Modernist architect Le Corbusier famously described the house as 'a machine for living in'. 'Smart' eco homes may be the ultimate expression of that idea, in that they automate processes of energy and resource collection to make sustainable living as easy, direct and efficient as possible. But that is only half of the equation. Nobody really wants to live inside a machine, no matter how efficient and comfortable it is. Although modern homes often have big picture windows to 'let the outside in', this really just gives the illusion of being in nature when we are actually at a remove from it. For our homes to truly be eco homes, they must give the people who live in them a deep sense of connection with the natural world, and the role they play in it.

A PART OF THE WORLD, NOT APART FROM IT

Our culture's current reliance on screen technology and private transport, as well as fragmenting work practices, puts us at risk of disconnection — not only from other people and our local community, but from nature itself. A connection to the land is part of our national identity. In a 2016 Colmar Brunton poll, New Zealand's natural beauty was the quality that Kiwis felt most proud of (90 per cent), with over two-thirds saying that caring for that environment is important to them.

It's a connection that traces back to our first people. Papatūānuku, the Earth Mother of Māori mythology, is the origin and nurturer of life, nourishing her children from her body (the Māori word 'whenua', meaning land, also means placenta). But whether born here, or an immigrant, every New Zealander has moments of connection with the land and water that are important to them.

My own childhood was one of summertime camping at Coromandel beaches, where I spent hours leaning over rock pools watching the busy activity of tiny hermit crabs, drifting transparent shrimps, darting guppies and jewel-coloured anemones. Today, when I'm stressed out, a few minutes staring into one of the tiny marine communities at the local beach is as good as a meditation session.

The great scientist and conservationist Edward O. Wilson describes this human affinity and longing for the natural world as 'biophilia' — a love for nature that is programmed into our DNA. A report commissioned for the Department of Conservation called 'The Nature of Wellbeing' looked at the way natural 'ecosystem services' contribute to New Zealanders' health and happiness. It described many tangible benefits we receive from nature, such as consuming fresh produce, the use of natural sources of energy and fuel, the wearing of natural fibres and the benefits we gain from plant-derived medicine, but also more intangible ones, such as the pleasure we take in leisurely bush walks and exciting nature-based adventure sports, opportunities to connect with others in our community over a shared love of and concern for nature, the inspiration we take from nature for acts of creativity, and a more equal society created by the sharing of free-use conservation territories.

To build an eco home is to ask ourselves: How do humans fit into the deeply interconnected web of relationships between plants and animals in this particular place on the planet? Bees are pollinators, worms are tillers of the soil, birds are dispersers of seeds, and fish are cleaners of algae. Even predators create balance in the system by ensuring that other species do not become too 'successful' for their own niche. And people, too, have their role. To me, it is to be midwives of change. Humans have an incredible ability to learn and bring new ideas into the world. However, our ability to invent new technologies is so advanced that it has disrupted the natural balance of the planet. In the big picture, this is normal — the planet has undergone constant global changes in its 4.5 billion years. But now that we see that our current systems are not working, it's our role to usher in new ones.

Some people think sustainable living is about saving the planet. This is quite the opposite of reality. The planet has survived for billions of years, and will continue to do so long after humans are gone. Sustainable living is about preserving our own quality of life, and making sure that our children, grandchildren and beyond can live as well or better than we do, with the same access to the dazzling diversity and richness of the natural world that we enjoy.

But the scale of the challenge of rethinking the way our homes are built can be intimidating, not to mention what we imagine the cost of making those changes might be. This book aims to take away the fear from that process — to give you an understanding of how sustainable design works, and how to make sustainable choices for your home that will work for you, your family and your budget, while improving your health and enabling you to experience the pleasure of connecting with nature.

CHAPTER 2

PRINCIPLES OF SUSTAINABLE BUILDING

Not every type of creature builds a home. Some are nomadic, some live in naturally occurring shelters like caves or trees — but for many species, including humans, the desire to build a home is inborn. Just cast an eye over the landscape of New Zealand, and it won't be long before you start to spy the elegant constructions of home-building creatures, both native and introduced.

In the fork of a tree you might spot a fine, tightly woven pīwakawaka (fantail) nest, lined with cobwebs and tapering to a point underneath, one of several a pair may build during a breeding season. Inside a decaying log, you might find the cellular honeycomb of a feral honeybee colony. Along the edges of streams, atop mānuka bushes, the white webs of the large native *Dolomedes minor* or nurseryweb spider glitter with dew on cool mornings. And on the hills of Central Otago, the pockmarks of rabbit holes are a familiar, if unwelcome, sight.

Even our national bird, the kiwi, whose name we borrow to describe ourselves, is an accomplished home-builder. Digging with its strong legs, it excavates burrows that it lines with leaves, grass and moss, covering the entrance with more sticks and leaves as camouflage. Some kiwi burrows are simple, little more than a scratched-out patch underneath the roots of a tree, but in the case of the great spotted kiwi, home can be a multiple-ended labyrinth of tunnels several metres long.

These homes are sustainable in a holistic sense. Once the inhabitant dies or moves on, the home either biodegrades or is reoccupied by another creature, either the same species or another, who might transform the home for its own purposes, like the bumblebees who build their nests of mud and wax 'pots' in abandoned rabbit burrows.

A BRIEF HISTORY OF HOMES IN NEW ZEALAND

To understand how to build or transform our homes into modern eco homes, we first need to understand what a typical Kiwi home is like, and how it came to be that way.

Here in New Zealand, our history of home-building is much shorter than in most countries. Our first human dwellings were wharepuni or 'sleeping houses', built by Māori from the 1300s onwards. With frames made from native timbers like tōtara and mānuka, and walls and roofs formed from raupō (bulrush), bark, moss and ponga (tree fern), these rectangular-shaped homes were our very first eco homes. They were constructed using natural and recyclable materials sourced locally, and designed with small windows and doors and earth-banked walls to keep them warm — although the ventilation did leave something to be desired.

When early Pākehā began arriving as whalers and sealers in the late

A century-old shepherd's cottage in Kaikōura was sustainably restored by Pattersons Architects, honouring the original home's native timber interiors and small windows, but improving its warmth and solidity. Photo: Simon Devitt

eighteenth century, with the help of local Māori, they integrated construction concepts from their homelands with local materials and Māori techniques, creating wood-framed clay and raupō houses, in a style that echoed European homes. These buildings were short-lived, lasting only a few years, and were cold and draughty. As time went on, settlers experimented with warmer and more solid forms of construction, including cob and turf houses, rammed earth, mud brick, and in the schist-rich Central Otago area, stone.

The interior of this workingman's cottage was restored by Pattersons Architects using authentic materials and natural finishes. Photo: Simon Devitt

By the late 1800s, the villas that are still common throughout the country began appearing, followed by California-style bungalows, with the designs ordered from international catalogues. Although beautiful and solidly built, often from native timbers such as kauri and rimu, these homes were draughty and cold in winter. In New Zealand's temperate coastal climate, humid air found its way inside easily, lingering and creating dampness and mould issues that persist in these buildings today.

With New Zealand's plentiful forestry resources, timber quickly became the main residential building material, although vast swathes of native forest were also clear-felled and burned for pastureland without logging, wasting a rich natural resource. From the 1900s, when it became apparent to the government that native forests wouldn't be a sustainable source of timber for construction, exotic plantation forests, mainly of *Pinus radiata*, were established.

As well as being cheap and plentiful, timber had the added benefit of performing better than masonry and brick during New Zealand's frequent, and sometimes severe, earthquakes. The Marlborough earthquake of 1848 and the Napier earthquake of 1931 caused widespread damage to both areas, prompting local architects and builders to review their approaches to design and construction. In Napier, the need for the city to be rebuilt in materials both earthquake- and fire-resistant (quake-related fires caused significant loss of life and property in wooden buildings) meant that young architects, energised by the international Art Deco movement, were given the opportunity to experiment with a new style of building, resulting in Napier becoming recognised as one of the great Art Deco centres of the world.

In 1936, the first Labour government in New Zealand established the Department of Housing Construction to manage their new state-housing scheme. Many of the early state houses were solid, simple cottages in a traditional English style, with small separate rooms, steep roofs with a small overhang, and tall casement windows — design choices better suited to the British climate than the New Zealand one. Due to government guidelines for home financing, most private homes of the 1940–60s were built in a similar way.

During these constrained decades, a handful of New Zealand architects began to work in earnest to define and create a style of modern architecture vernacular for New Zealand, one that reflected our unique landscape,

Many of the houses built in New Zealand in the early and mid-twentieth century were based on traditional English styles, using lightweight local construction materials. Photo: David Straight

climate, resources and cultural history. Vernon Brown, a London-born architect who moved to New Zealand at the age of 22, was at the forefront of this movement, developing a characteristic style of simple, informal lines inspired by the shape of sheds, a single-pitched roof with wide eaves, dark timber weatherboards, and large windows to embrace garden views.

These were homes designed to respond directly to the New Zealand context and environment, laying the groundwork for what would become modern sustainable home design. Although Brown and his similarly minded contemporaries created only a small number of private homes, from 1942 Brown taught at the Auckland School of Architecture, where he was able to pass on his ideas to a whole new generation of architects.

Through the 1970s and 1980s, rapid suburban expansion meant a proliferation of different home styles, again mainly copied from overseas cultures — from classic brick and tile, to Mediterranean, to colonial and 'contemporary', with open-plan layouts and greater indoor–outdoor connection. Concrete-block home construction, as used to great effect by Sir Miles Warren in Christchurch, and the Lockwood construction system, with its exposed knotty radiata pine interior, both became popular. Towards the end of the 1970s, the need for insulation in New Zealand homes finally achieved official recognition, with the introduction of minimum insulation levels in new houses.

With a growing international awareness of sustainable development and design as an answer to the concerns raised by the environmental movement of the 1970s and 1980s, the 1990s was a time of some progress, particularly with the passing of the landmark Resource Management Act in 1991, which enshrined in law sustainable management of our natural resources. The international Earth Summit in Rio in 1992 also focused global attention on the importance of preserving resources for future generations.

However, while these positive changes were underway, an unforeseen building crisis was brewing. In 1991, the New Zealand Building Act was passed, centralising building regulations at government level, and allowing increased self-regulation for the construction industry. Some builders and developers used this as an opportunity to cut corners. In 1995, a change to the New Zealand Standard for Timber Treatment meant that radiata pine that had not been chemically treated to resist rot could now be used for structural framing.

At the time, a bastardised 'Mediterranean' building aesthetic that minimised eaves and flashings, and used monolithic fibre cement or plaster cladding to mimic the look of stucco, masonry or earthen walls became popular. This cladding was often installed without space within the walls to allow moisture to evaporate, so when water made its way in (which it does in all buildings, but was more easily able to do through poorly protected monolithic claddings) some untreated timber developed mould and rot issues that were disastrous for the health of the homeowners and the stability

The 'Superhome' in Christchurch, designed by Bob Burnett Architecture, was the first home to achieve a 10-star rating from Homestar. It incorporates more than 20 innovative sustainable building techniques. Photo: Juliet Nicholas

of the buildings. The 'leaky homes crisis', as it came to be known, affected over 100,000 buildings and cost more than $30 billion in reparations. Two decades on, some properties are still affected.

In 2005, the New Zealand Green Building Council (NZGBC) was established to support and encourage more sustainable and environmentally friendly building practices in New Zealand. Initially, they focused on commercial buildings, introducing the Green Star system in 2007 to rate the environmental and health standards of commercial buildings. This system was based on the Australian Green Star system, which in turn followed well-established international 'green building' rating systems like the American LEED (Leadership in Energy and Environmental Design) certification and the British Code for Sustainable Homes.

In 2010, the NZGBC launched the Homestar system (see sidebar on page 29), which provides an easily understandable measure of a home's 'green' credentials. Standalone homes, terraced homes and apartments are rated separately, and both new builds and renovation projects can be assessed.

Today, more than 20,000 houses are registered to receive Homestar ratings, and sustainable design is, if not exactly mainstream, at least being built into the processes of more architects and designers, and is on the map for developers and the construction industry. Many architects informally report that interest in sustainable design varies widely from client to client, with some seeing it as a guiding light, but many others putting it in the

A small natural stream ran through the East Auckland coastal property this house was built on. During construction, the natural watercourse was recreated by Daniel Marshall Architects to become a beautiful feature tying the house and landscape together. Photos: Simon Devitt

category of 'nice to have, but not a priority'. In terms of government-led building standards, New Zealand lags badly behind the rest of the world; Andrew Eagles from the NZGBC says that, unlike most OECD countries, we don't have energy-performance certificates for newly built homes. By 2021, all of Europe will be building to zero or near-zero energy standards for homes. That puts us about a decade behind international standards, a situation Eagles describes as 'woeful'.

At the most informed end of the spectrum, homeowners, architects, designers and builders understand that every home is an ecosystem of its own, one that integrates with the grander ecosystem, first in a local context, but also at a national and even international scale. An eco home that has, for example, a solar-power system generating its own electricity from a free, renewable source, has immediate and direct benefits to the comfort, health and wallets of the people who live in the home. The reduced demand on electricity from non-renewable sources elsewhere in the country moves the country as a whole closer to 100 per cent renewable energy generation. And the country's transition away from use of fossil fuels decreases greenhouse-gas generation, which functions at a global level.

Building or renovating in a sustainable way is not about one particular 'best' choice in materials, layout, systems or design approach. Sustainable design is a holistic set of principles that can be applied in countless different ways in response to the needs and capacities of your location, family, budget and lifestyle. An eco home for an older couple without children, living in Auckland and working two jobs in the central city will be quite different from an eco home for a young family of five living on a semi-rural property on the fringe of Christchurch with one person working from home most of the time.

At its most basic, eco-home design starts by asking the question — to paraphrase Jason F. McLennan's manifesto *The Philosophy of Sustainable Design* — how can we maximise the quality of the built environment while minimising or eliminating negative effects on the natural environment? Broadly, it encompasses choices around where the home is built, how the home is oriented and designed, and what materials the home is constructed from.

SIX PRINCIPLES FOR ECO-HOME DESIGN

1. RESPECT THE SITE AND ITS COMMUNITY

In his iconic 1949 book *A Sand County Almanac*, the American environmentalist Aldo Leopold wrote of the 'land ethic' — the understanding that people are inhabitants of the earth no more or less valuable than any other creature or plant species, and that every living thing has interconnected needs and rights. Nature cannot be compartmentalised; what we do on our own piece of land affects the wider ecosystem. The most current example of this is our degraded rivers, lakes and streams and even city bays, polluted by run-off from thousands of farms, factories, and the chemical sludge washed into stormwater drains from urban roads, driveways and roofs.

Any true eco home will be planned with the starting point that buildings are intrusions into existing landscapes, so they must preserve, accommodate and nurture the habitat they become a part of. This means paying close attention to the flora and fauna native to a site and its surroundings, the climate, the geographical characteristics of the area, including water flows, and the human community around it.

This is not about protecting individual creatures or plants so much as preserving environments where biodiversity can flourish. Our planet is a magnificently complex and diverse ecosystem, where every change has ripple effects. When one element of a habitat is stressed or destroyed, the other elements of the habitat are weakened and more prone to collapse. The mind-bending Haast's eagle, a gigantic terror of the pre-1500s South Island, disappeared from our skies not because it had any natural predators, but because its main food source, the giant moa, was hunted to extinction. The loss of one species is the starting point for a chain reaction of losses.

As an isolated island, New Zealand has evolved a rich and unique indigenous biodiversity, full of plants and creatures found nowhere else in the world. But despite our short human history, our biodiversity has been destroyed to a shocking degree. The Department of Conservation reports that almost a third of our indigenous birds have been exterminated by humans and introduced pests. More than 63 per cent of our original forest has been converted to farmland, urban land and roads, destroying and shrinking the habitats of thousands of species. Many of our surviving iconic creatures have been driven to the edge of extinction. The little spotted kiwi, once found throughout the country, is now limited to a smattering of protected reserves and offshore islands. Tuatara — literally a relic of a previous age, and nothing if not survivors — have likewise been driven to extinction on the main islands and exist only on small offshore islands.

These creatures exist at the most visible end of a phenomenally complex ecosystem of vertebrates, invertebrates, plants, soil, water, fungi and bacteria that works as a whole to support life in New Zealand. Every one of us, from farmers to city dwellers, plays a part in preserving this irreplaceable natural habitat, which underpins billions of dollars of annual GDP, not to mention the sense of shared national identity and culture we gain from it.

In a practical sense, this must be addressed first at a local and an individual level. Every owner of a piece of land has a responsibility to identify, preserve and protect the distinctive natural community and resources that belong to it.

HOW DOES HOMESTAR WORK?

The Homestar system rates new and substantially renovated homes on a scale, with 6 being a home that is substantially better than the Building Code standard, and 10 being a 'superhome' in terms of health, comfort and environmental performance. Most homes built before the turn of the millennium would be lucky to scrape through the system with a score of 2 or 3. Surprisingly, even a new home built to the minimum standards of the New Zealand Building Code would only achieve a 4-star rating.

Building to a higher standard — a 6 or 7 is recommended by the NZGBC — does cost a little more — around 1.5 per cent, according to an independent Homestar Cost Benefit Analysis performed in 2013. However, it pays itself back with annual energy and water savings of around $800 (for warmer areas like Auckland and Northland) to $1000 (for cooler southern areas) every year.

When your home is assessed by a Homestar professional, they'll measure it against a set of environmental, health and energy consumption factors. To reach a Homestar 6-star rating, homes must achieve a certain number of points across a range of features, including better-than-minimum insulation levels, efficient energy and water use, extraction fans vented to the outside, LED or CFL lights, an outside washing line or condensing dryer, certain standards for shading and ventilation, and using products certified by Environmental Choice or other recognised certifications. Even aspects like close access to transport services, surface water run-off and fruit trees in the garden can earn points towards the rating.

While the Homestar rating system doesn't have performance standards quite as rigorous as those for certified Passive Houses or Living Buildings (see descriptions later in the chapter), it does evaluate a wider range of features, meaning it works particularly well when paired with other certifications. The system aims to 'reward better building', says CEO Andrew Eagles. Their big goal, he says, is to raise the standards of building across the market — something that extremely high-spec certifications like Passive House have a limited ability to achieve.

Owners of existing homes that wouldn't be likely to receive a Homestar rating of 6 or higher can have their home assessed for the new HomeFit or HomeFit Plus standards, launched in May 2018, which offer a basic 'warrant of fitness' for a home's heating, ventilation, hot-water system, smoke alarms and insulation in a 35–40 minute check.

2. USE RENEWABLE NATURAL RESOURCES EFFICIENTLY

Sustainable home design makes optimal use of resources that are free and endlessly renewable — such as light and heat from the sun, wind for ventilation and power, and rainwater harvesting. By building in a way that draws first on natural heat, light, air and water movement and rainfall, we reduce — or in some cases eliminate — our need for energy derived from fossil fuels, our load on the national power and water grids, and our use of refrigerants for cooling air, while at the same time lightening the load on our wallets.

This is achieved through starting with good 'passive design', or orienting the home on the site to let the sun shine in when the house needs warming, using eaves or other shading devices to protect the areas of the house that are at risk of overheating when the sun is at its strongest, and planning room layouts to work well with the passage of the sun through the day.

Passive design may also incorporate the use of thermal mass, or large areas of dense, solid materials like concrete, brick or masonry that can absorb and store the sun's heat to slowly release it later in the day, keeping the indoor temperature stable. Good insulation is also important to keep the heat inside once it's there. Windows are positioned to allow optimum natural light to pass through the house, reducing the need for artificial lighting, and making sure the windows can be opened in a way that encourages natural cross-ventilation, to allow cool, fresh air to move through the house. Bodies of water, such as pools or fishponds, can also be used to help cool the air that moves across them and into the house.

All these aspects work together, so should be considered holistically, with the most benefits coming when they're incorporated as part of a new build. Apart from the small cost of specifying higher-than-standard insulation levels, incorporating good passive design into a home does not make building your home any more expensive, and will save you considerable money in reduced running costs over the long term.

On top of passive-design elements, renewable resources can be collected through active systems like solar-power electricity or water heating, wind power, and rainwater collection and filtering. These systems do need an upfront investment and some care and maintenance, but can meet up to 100 per cent of a home's electricity and water costs, saving money over the lifetime of the system.

An extra benefit of collecting and using renewable resources directly is that it strengthens our understanding of and connection to how nature works, and heightens our awareness of our true levels of consumption. Many people have no idea how much electricity it takes to run a drier or a refrigerator for an hour, nor how much sunshine is required to produce that electricity. When you have your own solar-power system, that information is right at your fingertips.

This family home in the Wairarapa by Donnell & Day uses skylights and large windows to make the most of winter sun, heating the concrete-slab floor and reducing the occupants' reliance on electricity, even on the coldest of days. Photos: Duncan Innes

3. CHOOSE LOW-IMPACT MATERIALS

Buildings are complex systems that need to perform and please in many different ways, so when it comes to materials, there are few absolute right or wrong choices. Sustainably harvested local timber may, at a basic level, be a 'greener' material than concrete, but a concrete floor performs far better than a wooden one as a form of thermal mass to store free heat from the sun. But while concrete flooring may be a superior form of thermal mass for your home, if you hate the look and feel of it, you won't enjoy living in your home.

What you want from a material is that it is, first, fit for its purpose; second, that it ticks as many boxes as possible for being non-toxic, sustainably sourced, and recyclable or biodegradable at the end of its (ideally long) life; and third, that it's aesthetically pleasing to you.

Life-cycle analysis (LCA) is a way of looking at materials from 'cradle to grave', by evaluating how they are made, transported, used, recycled and disposed of, with the aim of making more informed choices about their environmental friendliness.

Some good LCA questions to ask when considering material choices are:

- Where does this material come from — how is it extracted, harvested or produced, and what sort of waste, by-products, energy use and emissions are part of that process?
- Who are the workers who extracted, harvested or produced the material, and were they treated well?
- How does the material get to my building site — does it travel a long way or a short way, and what form of transport does it take?
- Does it need to be packaged in lots of other materials to safely get to the site, and what happens to those materials afterwards?
- How much of the product is wasted in the installation process, and can any waste be effectively reused or recycled?
- Does the finished product release any harmful by-products such as volatile organic compounds (VOCs) after it has been installed, and if so, how long does it do so for?
- How long will the product last, and what kind of maintenance will be needed to keep it in good shape?
- At the end of this building's lifetime, can the material be easily separated from other materials and recycled, reused or biodegraded easily?

WHAT IS CO-HOUSING?

Described by Cohousing New Zealand as 'intentional clustered housing with some common facilities and shared neighbourhood life', you can think of co-housing as a deliberate community where the people who intend to own and live in the houses get to contribute jointly to all decisions around the design and running of their neighbourhood. The members of the community often share aspects of daily life, like group meals in a common house, fruit orchards and vegetable gardens, and also pool resources like tools and vehicles.

The concept was born in Denmark in the 1970s, where 10 per cent of homes are now in co-housing communities. It's also found favour in the United States and Europe, and there are nascent co-housing models in Australia, the United Kingdom, Canada and even in New Zealand.

New Zealand's first co-housing development was the Earthsong Eco-Neighbourhood in West Auckland, which began as a project in 1995, with the first homes being completed in 2001. Today it has more than 30 homes, and serves as inspiration for at least 10 more co-housing projects at various stages throughout the country.

The next co-housing project set for completion is the High Street Cohousing Project in Dunedin, which began its journey in late 2013. Future resident Catherine Spencer believes co-housing is the only sensible way to live. Decisions are made using the non-hierarchical 'coloured card consensus' decision-making process, which Spencer describes as a creative, flexible, solution-based way of working.

Challenges have come in the form of establishing a good legal structure for the co-housing project and securing finance, as the makeup of the community changes over time in the lead-up to breaking ground.

The Dunedin High Street Cohousing Project, designed by Architype (whose director, Tim Ross, is a future resident) clusters a variety of Passive House-standard homes (see sidebar on page 34) around a shared green space, with a large community house where residents can meet, share meals and accommodate guests. Image: Architype

WHAT IS A 'PASSIVE HOUSE'?

Passive House (also known as PassivHaus) certification is a standard for extremely low-energy buildings that was first developed in Germany towards the end of the 1980s. Its aim is to create warm, dry, comfortable buildings, while using as little electricity as possible. Although the certification is called Passive House, any type of building can be built to Passive House standards.

A typical Passive House has energy needs that are 90 per cent lower than the average existing New Zealand home, and about 75 per cent lower than a new house built to Building Code standard. That means that a solar-power system, which will normally only supply around 25–30 per cent of an average household's power needs, will probably be capable of servicing 100 per cent of a Passive House's power needs.

Passive House certification is voluntary, but very strict. The key focuses are good passive design — the best possible site orientation, massing and shading — and the house's 'thermal envelope' and airtightness. Architects and designers achieve the high standards required through detailed thermal modelling techniques and design strategies from the Passive House Planning Package.

Since Passive House certification is a performance standard, not a design system, there are no rules around what materials the house must (or must not) contain, nor the style of architecture. You could build a Passive House that looks like a traditional weatherboard villa, or one with a very contemporary shape and monolithic façade.

In general, Passive Houses work better as smaller and simpler forms, because larger and more complex forms require more insulation and detailing, making it more difficult to achieve the performance standards. You can build bigger and more complex Passive Houses, but, in short, it will cost more money. That said, the approach can be a cost-effective way of building.

Passive Houses have been a little slow to take off here, as architects, builders and councils have lacked experience in Passive House construction methods and how they line up with our Building Code solutions and standards. The first New Zealand Passive House was completed in Auckland in 2012, and by the end of 2017 there were still only 19 completed and certified Passive Houses in New Zealand. However, a number of architects and designers are integrating Passive House principles and techniques into their work without aiming for full certification, so awareness of the typology is on the way up.

The George Street Passive House in Wanaka was the first Passive House completed in the South Island. Designed by Rafe Maclean, it includes triple-glazed windows, super-insulation and a mechanical ventilation system. It needs only a small heater to stay warm. Photos: Simon Devitt

4. MOVE FROM WASTE TO REGENERATION

One of the biggest challenges of moving to a more sustainable lifestyle is accepting that we are part of a culture of gross wastage and overconsumption — even those of us who think of ourselves as environmentally conscious. Our society's focus on constant consumption, an endless upgrading from 'less to more' and from 'more to better' can mean that buying things feels like a birthright. It can be hard to accept the idea of buying less, because it feels like depriving ourselves in some way.

You've probably heard the mantra 'Reduce, Reuse, Recycle'. What you may not realise is that the saying works in order of importance. When it comes to minimising environmental impact, the best thing you can do is first reduce the amount you buy, then reuse what you already have (perhaps in a different way to its first use), and finally recycle anything that has come to the end of its life.

Take a walk through the nearest big-box warehouse store, where long aisles of shelves are stacked beyond human height with thousands upon thousands of items — decorative photo frames, frypans, plastic action figures based on this year's hit Disney characters, hose fittings, novelty beer mugs, slogan T-shirts, beach umbrellas, party decorations, fan heaters, pens, coat-hooks, scrubbing brushes, candleholders, drawer organisers — anything you can imagine needing or wanting for your home or person.

Now consider that nearly every item in the store will still exist long after you're dead. They may not be in use — in fact, they almost certainly won't be, given that the average lifespan for even large, complex goods like televisions is less than six years. Those plastic party decorations may be crumpled in the bin 48 hours after they've been bought, but they will linger on in landfill for hundreds of years. So will millions of other items sold in stores, supermarkets, boutiques and malls up and down the country every year.

Recycling, while an essential part of sustainable living, is not a full solution as it currently exists. You can't pat yourself on the back and call yourself an environmentalist for recycling — that's just the lowest bar. In addition, recycled materials for construction often need to be treated, refinished or completely overhauled to make them useful again. As architect Nicola Herbst says, 'There is a premium on recycled material. Someone's got to find it and sift through it and get it into a state that's usable from an architectural and engineering perspective. It's not just about going and getting some timber off-cuts.'

Most modern recycling is actually 'downcycling', not true recycling, where a material can be endlessly returned to its original form. When you throw a plastic milk bottle into the recycling bin, it's unlikely to have another life as a milk bottle, because it's difficult and expensive to return recycled plastic to a food-grade quality, especially after it has been recycled more than once. Instead, after being crushed and bound into bales, and shipped to an overseas recycling plant, probably in South-East

Asia, accumulating carbon kilometres as it goes, it may end up as carpet underlay, polyester fabric, shoe soles, kids' toys or a road cone. At the end of that use, it may well end up in landfill after all.

Every year, New Zealanders send around 2.5 million tonnes of waste to landfill, or around 530 kg of waste per person. And every time a home is built, on average, 4 tonnes of construction waste goes to landfill. Yet the Building Research Association of New Zealand (BRANZ) estimates that at least half of this construction waste could be diverted just with better sorting on site.

The reality is that much waste can be avoided simply by planning well in the first place. Items that we throw away repeatedly and replace (like toothbrushes) can be replaced with substitutes with much more longevity or bought in bulk rather than single-serves. There are many creative ways to decrease consumption without creating a sense of deprivation, such as gifting experiences rather than products.

Another kind of wastage occurs when we knock down old buildings and build new ones in their place. Although some materials can be salvaged from demolition, in most cases new materials are used. The manufacturing of these new materials uses a large amount of energy and produces greenhouse gases and pollutants. Although new buildings are usually more energy-efficient than old ones, it may still take decades to achieve payback on the embodied energy costs of making new buildings. With immediate action to reduce climate change being at a critical point, this means it may be better for the environment to improve old 'bad' buildings than demolish them and build new 'good' ones.

Ultimately, as architect William McDonough and chemist Michael Braungart wrote in their revolutionary book *Cradle to Cradle: Remaking the Way We Make Things*, the future is in looking to nature to learn how to redesign our systems to make them endlessly abundant. In the same way that the waste from our tables can be turned into compost to fertilise our vegetable gardens to grow more food for our tables, the future of man-made materials is in learning how to improve their recyclability to the point where they can be reused endlessly.

Imagine, say McDonough and Braungart, if instead of buying a new television every six years, you simply paid an annual hire fee to a television manufacturer, who claims back your television every six years and replaces it with the newest model. The manufacturer then breaks down your old one, retrieves all the valuable materials, and recycles them into new televisions. What was once waste destined for landfill becomes a kind of synthetic 'nutrient' that operates in a continuous loop, exactly like a natural system.

DEEP GREEN? LIGHT GREEN? BROWN-GREEN? A GUIDE TO CERTIFICATIONS

If it seems like there's an overwhelming list of considerations when choosing materials and products, don't worry — it's not just you. Responsible product and material choice is a difficult process to work through, even for professionals, not least because many products and companies have obscure chains of production and supply.

Others, too, indulge in a practice known as 'greenwashing': making vague but positive-sounding claims about their product (such as 'Contains natural materials!'), but not specifying clearly what or how much. Or focusing on one environmentally friendly aspect of the product ('Chemical free!') while ignoring less environmentally friendly ones, like a high embodied energy (the amount of energy used in extracting and manufacturing the product).

Some manufacturers add credibility to their products with certifications or badges. Some of these certifications are useful, and some are essentially meaningless. There are three main types of environmental certifications:

- **First-party certifications**, which are created by the company itself. This might consist of putting the words 'Natural' or 'Chemical-free' with a logo that looks like a tick or similar. This is a marketing technique to emphasise a positive attribute of the product (which may not tell the whole story) and hasn't been verified by an external party.

- **Second-party certificates** are given by the industry or trade associations. This may mean that the product is of a higher quality, but is not completely independent.

- **Third-party certificates** are awarded by independent organisations and verify that the claims the product manufacturer makes are true, or that the product achieves a certain standard of performance. This is the highest level of assurance about a product's environmental standards. The more commonly seen environmental certifications in New Zealand are listed below.

Declare (USA): Issued by the Living Future Institute, building products with a Declare label offer life-cycle transparency. The label clearly states a product's place of manufacture, materials/ingredients (including where they're from), VOC content, recyclability and life expectancy. Chemicals that are on Declare's Red List ('worst in class') are highlighted on the label.

Energy Rating Label (New Zealand and Australia): The trans-Tasman Energy Rating Label gives electrical appliances a 1–6 star rating based on energy efficiency, as well as a direct consumption figure in kilowatt hours per year, based on average usage.

Environmental Choice (New Zealand): This is our official environmental performance label, assessing products on multiple criteria relevant to our market. It's owned by the Ministry for the Environment, but operates independently. It is a member of the Global Ecolabelling Network, and has a mutual recognition agreement with all other labels in the network, including Germany's Blue Angel (Blauer Engel), the US's Green Seal and Singapore's Green Leaf. It evaluates, among others, building and construction products, flooring, paints, furniture and cleaning products. Homestar recognises Environmental Choice products in its ratings system.

Forest Stewardship Council (international): A non-profit organisation present in over 100 countries that certifies managed forestries and printers on their environmental and economic sustainability, and social responsibility.

Global GreenTag (international): A comprehensive independent product and materials rating system that's recognised in over 70 countries. It has four tag levels — Bronze, Silver, Gold and Platinum — that take a full life-cycle analysis approach, with Platinum classed as 'World-Leading'. The GreenRate tag indicates that the product complies with New Zealand's Building Code. GreenRate gives it a 'sustainability factor' rating of A (100 per cent), B (75 per cent) or C (50 per cent).

Greenguard (international): A certification from independent safety and standards-auditing company UL Environment, which evaluates building products, fittings and furniture for VOC emissions. Greenguard and Greenguard Gold certified products have passed rigorous low-emissions tests.

Programme for the Endorsement of Forest Certification (international): An alternative wood and paper certification scheme to FSC, this certification recognises localised certification schemes with a focus on each area's priority environmental and ethics issues.

Water Efficiency Labelling Scheme (New Zealand): The WELS sticker, issued through the Ministry for the Environment, gives new dishwashers, washing machines, toilets, showers and taps a 1–6 star water-efficiency rating, and as a direct water consumption in litres per minute/wash/flush.

WHAT IS A 'TINY HOUSE'?

More a social movement than an official building form, Tiny Houses have caught the imagination of people around the world (usually young singles and couples, or older empty-nesters) looking for a more affordable, minimal and low-impact way of living . . . and with the potential to up sticks and hit the road when you're ready for a change of scenery.

There are no formal rules for what makes a very small house a Tiny House, but, in general, they're under 40 sqm (less than a fifth of the size of an average new-build home in New Zealand) and they're usually transportable — whether that means they're fixed on a trailer (which can, at least temporarily, avoid the need to obtain a building consent for the house), or that they're designed to be relatively easy to move, like a shipping-container house.

A typical Tiny House costs anywhere from $50,000 to over $150,000 to build, depending on its specifications (and excluding the cost of land to put it on). There are a number of local companies producing standardised or customised Tiny House designs to suit all tastes.

In New Zealand, trailer-fixed Tiny Houses need to be sized under the New Zealand Transport Authority's limits for light trailers: less than 2.55 m wide, 4.3 m high, 12.5 m long and 3.5 tonnes. These can be parked on private land as temporary or permanent dwellings, and can be designed to be plumbed into the power, water and sewerage grids, or to be completely self-sufficient. District councils often apply the same rules to trailer-fixed Tiny Houses that they do to caravans, but this varies from council to council, so it's essential to check with yours.

Rules around how long a Tiny House can stay on a piece of land before being regarded as a permanent dwelling (potentially becoming subject to additional consents and requirements) vary across councils. Some subdivision developments also specify a minimum floor size for permanent dwellings, which a fixed-footing Tiny House would be unlikely to meet. If you're interested in a Tiny House, it's important to figure out exactly where and for how long you'd like to live in it first, and then get very clear on the local regulations.

This ultra-compact Tiny House, owned by Build Tiny founder Gina Stevens, incorporates a modern kitchen, living space, bathroom and laundry, sleeping loft and office loft — all on the back of a trailer. It's aimed at Millennials who are struggling to get on the housing ladder. Photos: Jemma Wells

5. DESIGN FOR A LIFETIME

The longer a building lasts in good condition, the better it is for the environment. And when you live in a good, comfortable, healthy house, you're more likely to stay there for longer, accruing all the benefits of long-term tenure. Selling and moving house incurs costs — transactional ones, such as real estate agent and lawyers' fees, mortgage-break fees and moving costs — but more importantly social ones, such as the disruption of children having to move schools, and disconnection from communities that we have established ourselves in.

When people move, it's often because their home no longer serves their needs — too small for a growing family or too expensive to maintain for empty-nesters; becoming too difficult to get around in as the owners are faced with the challenges of injury, illness or age-related disability.

Historically, New Zealand homes have not been great at serving the needs of people through more than a few years of their lives, and as a whole our housing stock is increasingly less well fitted for the needs of our changing and aging population. By the year 2061, there are expected to be more than 1.4 million people aged over 60 in New Zealand, with a large swell in the demographic reaching that age between 2040 and 2060.

So a typical three- to four-bedroom family home is probably not going to be well suited for older couples and singles, who often want smaller, warmer, lower maintenance houses, or at least homes where they can close off areas until needed for visiting family or friends. People with some sort of disability (one in four New Zealanders, according to the Office for Disability Issues) may also have to spend considerable amounts of their own and publically funded money adapting their homes to accommodate their level of ability. These changes can be as simple as ensuring level or accessible entry and exit points, bathrooms and kitchens — elements that can be easily designed into new homes.

When you're designing a home, it's tempting to design your own 'dream home' — an idiosyncratic tribute to your unique needs, family size and individual tastes. But the reality is that your home is almost certain to pass on to other people in the future, so taking a 'lifetime design' approach is more sustainable.

Considering how your home could be used flexibly or be relatively easily adapted, and how accessible the thresholds and entrances are (imagine whether a friend or family member in a wheelchair could come in and out easily) will give it greater longevity, and reduce the likelihood that in 40 or 50 years it will be knocked down and rebuilt or extensively (and expensively) changed to suit the needs of a future owner.

The Lifemark star system rates homes (from 1–5 stars) on how easy they are for people to live in for a lifetime. It considers factors like the width of carpark spaces, doorways and corridors; entrance accessibility; the height of power-points and location of appliances; the ease of handle use; and the ability for the home to be adapted easily in the future. There's an easy 'Homescore' tool at www.lifemark.co.nz to assess the basics of how your home would perform.

6. INTEGRATE A CONNECTION WITH NATURE

Since the 1970s, one of the terms most frequently used by real estate agents to advertise a desirable Kiwi home has been 'great indoor–outdoor flow'. Our (mostly) temperate climate and affection for the great outdoors means that newer homes are often designed with big ranch sliders or bifold doors in the living and bedroom areas to let us open the house up to gardens or patios.

But why stop there? The more that natural textures, colours, materials and other elements — particularly ones that are a beautiful reflection of the natural ecology of the local landscape — are brought into our homes, the deeper our sense of connection to nature becomes. The writer John Naisbitt, who studies future trends, predicts that as our lives increasingly use and even depend on digital technology, we will feel an even stronger pull towards human and natural connection to provide a sense of relief or grounding — a balance of 'high tech' and 'high touch'.

A good starting point is the use of natural local materials, as specifically local as possible. Not only does this reduce the distance that materials have to travel to the building site, it links your home with others in the area to create a distinctive community of homes, connected by their relationship to the natural resources of the area. Think of Central Otago, and the beautiful schist homes and buildings that characterise the area.

In his insightful book *The Shape of Green*, author Lance Hosey argues that in order for sustainable design to be enduring, it must be pleasing to look at, because we look after what we love, and we love what is beautiful. A home that is aesthetically delightful is one that will be cared for and maintained — an important aspect of building longevity, as many buildings are demolished for no reason other than that they have been allowed to deteriorate to the point where it would be more expensive to repair them than to build new, despite the environmental costs associated with new buildings.

WHAT IS A 'LIVING BUILDING'?

The Living Building Challenge aims to answer the question 'What does "good" look like?' Based on the principle of regenerative design — the idea that a building should give back more to the environment than it takes away — Living Buildings must perform to a set of rigorous criteria across seven 'petals': place, water, energy, health and happiness, materials, equity and beauty.

They must be completely self-sufficient — able to supply all their own water and energy requirements off-grid, and also able to account for all their own waste production, meaning that all wastewater (from toilets, washing, etc.) must be able to be treated on-site. All materials used in the building must be free of the Living Future Institute's 'Red List' chemicals, which are deemed to be the most environmentally damaging and toxic to human health. A sense of connection to the natural features of the site is part of the design standards, as is a design that is beautiful and healthy for the inhabitants.

An important aspect is that certification is only awarded after the house is built and the specified performance standards have been measured and met for 12 months. This means that a house can't just be designed to perform well on paper — it has to do it in real life as well.

It's a tough challenge to meet, but it's already been achieved in New Zealand, by the Te Kura Whare community centre in Tāneatua, Bay of Plenty, designed by Jasmax for the Ngāi Tūhoe iwi, and completed in late 2014. At least a dozen buildings are now registered in New Zealand for partial or full certification under the Living Building Challenge, including several Zero Energy homes and buildings (which derive 100 per cent of their ongoing energy needs from renewable resources), and a handful of residential projects aiming for full certification.

Designed by Collingridge and Smith Architects, the Living House in Beachlands will be 100 per cent self-sustained, including being Zero Energy, built to Passive House standards and treating all wastewater on site. Images: Abdallah Alayan

CHAPTER 3

CHOOSING YOUR PROPERTY

With clever design, a difficult site can be transformed into an eco home — like this steep, west-facing site in Wellington. Photo: David Straight

A flat, open, semi-rural site allowed for optimal house orientation for this Wairarapa home, as well as space for vegetable gardens. Photos: Elizabeth Goodall

Everyone knows the most important thing about a property: location, location, location. Or is it? In New Zealand, we have a habit of thinking about houses as financial investments as much as places to live in. That's one of the reasons driving our poor-quality housing — if maximising capital gain is your priority when buying a home, there's more incentive to invest in cheaper visual improvements like nice taps, new paint or re-carpeting than less visible but more substantial investments like good insulation and double-glazing. It also makes it tempting to buy a property for its potential to add value via a new kitchen, bathroom or living-room extension, rather than its comfort and sustainability through the seasons and years.

That approach sells ourselves and our families short. One of the main reasons people move is that the home they have doesn't meet their needs — it's too small, too uncomfortable, too far away from where they need to go regularly — and that can be a result of prioritising the wrong things at the time of purchase. Given the financial and social costs of relocating, and the health and comfort costs of a cold, damp home, the wisest investment you can make is time spent thinking and asking questions about *all* aspects of a site's or house's performance and potential before you commit to it.

If you're looking for a new place to live, you'll be considering one of three types of property: an empty site on which to build; a site with an existing property in poor condition to demolish and build new; or an existing home that may need renovation, or that will require maintenance in the future. If you're planning to create your dream eco home, there are some considerations that apply to all three types of property, and some specific to each type.

CONSIDERATIONS FOR ALL PROPERTIES

SERVICES AND COMMUNITY

Where you live has a huge effect on your day-to-day ability to live in a sustainable way. When you live within a 10-minute walk of public transport, shops, schools, parks and community facilities like libraries and churches, you decrease your reliance on your car while improving your health and engaging more deeply with your local community.

An online tool called Walkscore (www.walkscore.com) gives a rating out of 100 for the 'walkability' and proximity to public transport of any property, letting you see at a glance how likely you are to need a car to get around. But physical distance isn't the only thing to take into consideration — those 10-minute walks need to be well lit, safe for children and senior citizens, and ideally pleasant — shaded by trees or awnings, with accessible footpaths and road crossings. When considering a property or site, set aside 10 minutes before and after an open home or site visit to take a walk around the neighbourhood and get a feeling for it at ground level.

SUN, SHADE AND WIND

The sun is the ultimate free resource. A well-designed house on a well-oriented site will capture sunlight to store as heat (via thermal mass), energy (via solar panels) and light (through windows, skylights and solar tubes), reducing paid energy costs and making the house more pleasant to live in. However, it's also possible to have too much of a good thing; without proper shading, houses that get a lot of sunlight can overheat.

Large trees or tall neighbouring buildings can affect the amount of sun available to your home in both negative and positive ways — by either blocking useful sun, or protecting against sun that's too intense. Look for trees on-site that may need to be built around or removed (as a last resort), as well as young trees that will grow into much bigger ones. Also consider your neighbours' trees, as you will have little to no influence over whether they can be removed or pruned if they grow to block your sun.

In New Zealand, houses receive the most sun from a northern or north-eastern angle. When a property is described as 'north-facing', it means the windows and main living areas (or in the case of an empty site, clear space for living areas) are designed to face north. Since most urban sites are long and narrow, and Kiwis like to live in their backyards, sites where the back end of the property looks towards the north or north-east are often preferred. If the property you're considering has a lovely view, check which direction it faces. Southward views are harder to design around.

Although north-facing is best, an east–west orientation isn't the end of the world; east-facing living areas will get morning sun, and west-facing ones will catch the afternoon sun. Shading will be important for these angles, to optimise solar gain in winter and prevent overheating in summer. In winter, the sun shines in from a lower angle than in summer. When eaves and overhangs have been designed to the right depth, they will block out the hottest summer sun, while letting lower winter sun in through windows.

The New Zealand Institute for Water and Atmospheric research (NIWA) has a free tool called SolarView (available at solarview.niwa.co.nz) that will help you check the amount of sunlight a property receives throughout the day and year. This can be particularly helpful if you're thinking about putting in a solar-power system.

Prevailing winds can likewise be both useful and problematic. In summer, wind blowing through your home helps to cool and ventilate it naturally, especially if windows can be placed facing the prevailing wind and on the opposite side of the room to allow cross-ventilation. Since winds often change direction seasonally, the same windows that let in summer breezes may well face away from cold winter winds. Outdoor living areas should be placed away from prevailing winds if possible, or sheltered by landscaping. To get a sense of the prevailing wind direction and strength, walk the site over different days, look at the surrounding topography (valleys can create wind tunnels), and check nearby properties for landscaped windbreaks.

Architect Tim Gittos wanted a site close to the city to increase cyclability and walkability. One of the only available sites in his budget was a steeply sloped west-facing 'cheap and nasty' 'brownfields' plot in a 1960s subdivision. By pairing up with another couple to build two houses at once, they were able to effectively utilise an otherwise marginal piece of land. Photo: David Straight

SLOPE AND RUN-OFF

In general, the steeper the gradient of a site, the more difficult and costly it will be to build on it, thanks to the structural engineering and site-preparation costs. Design-and-build companies that offer standardised home models tend to prefer sites that are completely flat (sometimes described as 'builder's flat') to avoid design changes and additional costs.

A steeply sloped site may restrict the kind of foundation your house can have, and an existing home on a steep site will likely have a higher cost of rebuild for insurance than a similar home on a flatter site.

That said, sloped sites can often offer better views than flat ones, and a gentle gradient of around 10 degrees (which may appear flat to the eye if you're not a builder) can offer a natural direction of run-off. With retaining, a steeper slope of 20 degrees can accommodate a walk-out basement (a partially buried basement with a door to the outside, rather than a fully underground basement) without too much extra cost. Steep north-facing sites can offer excellent solar access, especially further south of the country, where the risk of overheating is lower than in the north.

Take a walk around the site and look for channels or sodden patches where water runs during heavy rain. If the property is below street level or other properties, water may flow into it from other properties. A Land Information Memorandum, available from the local council (or some real estate agents offer them to interested buyers) will indicate any flood plains on the property. Check for rainwater-collection systems, or whether there is room to put one in.

NEW ZEALAND'S CLIMATE AND TOPOGRAPHY

New Zealand's climate is classed as temperate, with four distinct seasons, but this generalisation doesn't tell the whole story. Our country's long, thin shape and mountainous backbone, high proportion of coastline to land area, and oceanic location gives us a complex range of sub-climates and a tendency towards dramatic shifts in weather. In the humid Far North, the climate tends more towards sub-tropical in the warmer months, while in the semi-arid Central Otago alpine region in the rain shadow of the Southern Alps, snowfall can reach ground level. On the other side of the Alps, the West Coast experiences the highest levels of rainfall, as north-west marine weather systems dump their moisture before passing across the mountains.

Under the New Zealand Building Code, the country is divided into three climate zones. Zone 1 includes Northland, Auckland, Thames-Coromandel and the Kermadec Islands. Zone 2 covers from the Waikato to East Cape, and down to Wellington, but excludes the central alpine area. Zone 3 is the North Island central plateau, the South Island, Stewart Island, and the Chatham Islands.

Each zone has different standards for thermal performance or minimum insulation levels in roofs, walls, floors, glazing and skylights in new homes. Zones 1 and 2 are very similar, while the colder Zone 3 requires higher standards. Remember that these standards are legal minimums only, and that houses built to the Building Code standard will not necessarily be comfortable to live in year-round without high levels of heating or cooling.

The Building Code also specifies zones for earthquake risk, wind, snow load and exposure or corrosion, which have associated design standards. Buildings in high-wind zones, for example, may require extra bracing, and coastal properties may need claddings that are better able to resist corrosion from salt-laced air.

The K Valley House on the Coromandel Peninsula respects the agricultural history of the area in its utilitarian design. The clients had two potential building sites in mind, but, after walking the property with architects Lance and Nicola Herbst, were convinced that a third spot was the best. Photo: Patrick Reynolds

USEFUL DOCUMENTS

Certificate of Title: This shows the size and shape of a section, rights of way, mortgages, leases and other interests on the property.

Land Information Memorandum (LIM): This shows the site features, such as erosion or flooding zones, stormwater and sewage, previously issued permits and building consents, Historic Places protections, zoning and permitted land uses, notices to and by the local council, and even rates owing.

Building inspection: A building inspector will examine a house for defects that need to be addressed immediately, and advise on any significant future maintenance requirements. The quality of building inspections can vary as they are unregulated, so ask family and friends who have purchased homes for recommendations.

UNCONVENTIONAL IDEA: BUILD ON THE UGLIEST PART OF THE SITE

In the 1977 architectural and town-planning handbook *A Pattern Language*, the authors suggest that rather than building on or near the flattest and most lovely part of a site, a more ecological choice is to build on a steep, ugly or neglected area. This leaves the most beautiful natural aspects of the site intact, and improves the ugliest and least useful areas with a building. Although flat land is the easiest to build on, it's also the best growing land. If having a productive garden as part of your site is important to you, this approach may give you the best of both worlds.

SOIL

Breaking ground is a momentous occasion in the timeline of a new home, so you don't want it to come with any unpleasant surprises (though a few surprises are usually par for the course). Ground stability is crucial for a successful build, and can affect the long-term maintenance costs of a home, so make sure you know that the property is stable, is not prone to erosion, liquefaction or slips, and won't settle.

If you are looking at a section in a planned subdivision, request a copy of the engineer's report from the developer. Look into the previous use of the land, to see whether there might be chemical residues from previous use as agricultural or industrial land, or subsidence from previous use as landfill.

Types of soil vary across the country, from dense, claggy clays that can become compacted and hardened by construction work, resulting in more surface run-off, to rich loams and fertile volcanic soils that are excellent for growing and drain well.

CONSIDERATIONS FOR UNBUILT SECTIONS

The empty urban plot is unicorn-like these days, though sections within planned developments, and unbuilt rural or coastal sites are more common. Sites outside city limits may have more flexibility in council restrictions towards on-site water treatment systems, but it's important to check local regulations, as they vary from district to district.

If you are buying a site within a planned development or subdivision, make sure you are aware of any building and material covenants on new builds. Some estates require a minimum house size (ruling out Tiny

Houses, or even substantially smaller-than-usual homes), restrictions on roofing and cladding materials, and even house colours. They may also include restrictions on the placement of solar-power systems — some subdivisions specify that they must not be visible from the road, or reflect glare towards neighbours.

Check that all the services you may need — power, gas, sewerage, water, fibre broadband — are available in the area and have been connected to the section. If it's a big site, and the service connections are a long way from where you want to build, it can be very expensive to bring them through. Also think about the vehicle access to the property — not just for your car, but for earth-moving equipment, delivery trucks and cranes. The closer you can get your deliveries to the building site, the lower your labour costs will be. If you are considering prefabricated or modular building, good access is essential.

CONSIDERATIONS FOR BUILT SITES FOR DEMOLITION

Urban land is increasingly hard to come by, and you may find yourself in the position of finding a perfectly located site that already has a run-down house on it. In the first instance, it's a good idea to ask a trusted builder or architect if they can see ways for the building to be redesigned or improved instead of demolished — in many cases, even a very extensive refurbishment is cheaper than building new, and excellent results can be achieved.

If the existing building truly can't be redeveloped to suit your needs, the next step is to conduct a pre-demolition audit to see what materials can be reclaimed and recycled. Some materials, such as timber or metal beams, roofing iron, garage doors, windows, doors and kitchen and bathroom sinks and laundry tubs can be sent to salvage yards. Bricks can repurposed for landscaping walls or an outdoor fireplace on-site, and old wool or natural-fibre carpets turned into weed matting. Fixed appliances in good working condition can be donated to charity shops or kept for the next council inorganic collection.

CONSIDERATIONS FOR FUTURE RENOVATIONS

New Zealand's housing shortage has led to a level of frantic activity and competition over the last few years that makes assessing existing houses increasingly difficult. If you're looking in a high-demand area, you can often find yourself in the insane position of feeling pressured to make an offer after just a couple of short visits to a house that you may be committing to for a decade or more. With many of New Zealand's houses built before 2004, when the current version of the Building Act came into force, chances are good that you'll need or want to make changes to improve or maintain a 'second-hand' home's condition. The more fundamental these changes are, the more they will cost.

When house-shopping, it's easy to fall in love with the look or feeling of a home and forget to pay attention to how the house functions throughout the days and seasons, and how much it will cost to change or maintain. New Zealand's busiest months for house sales are September, October, November, February and March — in other words, the warmer months (excluding the Christmas holiday period). Cold, damp homes become much harder to spot during these months.

Writing a list of details to check and questions that you need answered by the real estate agent will help you keep your head straight when you're starry-eyed about a home's beautiful character or wonderful view. Real estate agents aren't required to voluntarily disclose any negative features of a property, but they can't mislead you either. So ask, ask, ask about anything that's important to you.

Once you've checked out the important features of the location and site topography covered earlier, take a careful look at the structure of the house and its built-in systems. The details of each of these areas are covered in later chapters of this book.

EXTERIOR

- What sort of cladding/s and roof covering does the house have, and what condition are they in? Will they need maintenance like painting, cleaning or staining? If the cladding does need maintenance, will you need scaffolding to do it?

- What year was the house built? If the house has a monolithic cladding and was built in the 1990s or early 2000s up to 2003, has it been tested for weather tightness?

ENTRANCEWAYS

- How easy would the house be to access if a member of the household were to develop mobility issues? If it's not currently accessible, would it be easy to adapt?

FOUNDATION

- Does the house have a timber or concrete-slab floor? Is there underfloor insulation? If the house is an older villa or bungalow with

This house in Auckland was originally a run-down building on a great waterfront site. The clients were keen to renovate instead of demolish. Dorrington Atcheson Architects designed a dramatic and modern zinc-clad extension to the rear, with transitional timber recladding, resulting in a 'rich layering of materials'. Photos: Emma-Jane Hetherington

exposed wooden floors, lie down and see if you can see the ground beneath the house through the cracks between the floorboards. If you can, it means there's no floor insulation.

- If the house has a concrete floor or solid walls, are they being used as thermal mass to collect heat from sun falling on them? If the concrete slab is carpeted, could the carpets be removed to use the floor as thermal mass? Could windows be enlarged or moved to make better use of this?

INSULATION

- Ask the real estate agent whether the house is insulated in the walls, floor and ceiling, what type it is, and when it was installed.

VENTILATION

- What sort of windows does the house have, and are they easy to open and close? Do they have security stays so they can be left open to air or cool the house? Are insect screens fitted on any of the doors or windows?

- Look in the bathroom, kitchen and laundry for extraction fans and range hoods or bench-top extraction. Switch them on and off to check they work, and look on the outside of the house to make sure the fans vent to the outside, not the roof space.

- If the house has a Heat Recovery Ventilation system, ask the agent if it's a balanced-pressure system (draws air from outside) or a positive-pressure system (draws air from the roof space). Balanced-pressure systems are preferable, particularly in the South Island.

APPLIANCES AND FITTINGS

- Photograph and note the brand names of any appliances and fittings (like toilets, taps and showerheads) that will be left as chattels for the house, and check their energy- and water-efficiency ratings online later.

CHAPTER 4

BUILDING YOUR TEAM

Despite New Zealand's give-it-a-go DIY culture, building or substantially improving your home is a huge and complex undertaking that almost always requires the help of professionals, particularly if you are trying to achieve the performance standards of a modern eco home or work with a less mainstream approach or products.

The people you decide to work with will be your sounding boards and guides, helping you realise your big vision for your home (and gently bringing you back down to earth at times), and providing their own inspiration, expertise and ideas to add to yours. It's essential that you get on well with them, as you'll undoubtedly need to get through a few difficult conversations when things go wrong, budgets blow out, timelines drag on and decisions have to be made — or changed. And you need to be able to trust that they're on board with achieving what's important to you, not what's easiest for them.

DEVELOPING YOUR BRIEF

Having a clear and detailed brief to explain your eco home vision to the people you're working with will make the process of achieving it much easier. So the first person to have a conversation with is yourself — and the other people who will live in your home.

Start with the patterns of your daily lives. Think about how many people need rooms for sleeping and privacy; where you like to eat your meals day-to-day; where you like to relax; where you can go to work or study in peace. We live our lives in surprisingly habitual ways. Identify the routines or habits that add pleasure to your life (or that you would like to develop), like having a morning coffee in the sun, a weekly soak in the bath, lying down in a warm or cool spot to read a book or browse the internet, having lots of friends over for a meal.

Next, add the functional parts. Consider how many bathrooms will make life pleasant for everyone, how big the kitchen needs to be considering how much you like to cook and entertain (or not), how much washing you do, and how much storage you need — both for everyday household stuff and for big items like cars, bikes, boats, surfboards and tool collections.

Then — since you're building an eco home — define the aspects of eco-living that are most important to you. Do you want your home to be 'off-grid', or as close as possible within local regulations? Do you want your home to have a productive and beautiful garden? Is it important to you that the materials used in your house are as natural and non-toxic as possible? Do you want your home to be carbon-neutral? Do you have a particular Homestar rating or design approach, such as Passive House, in mind? Although you may want to include all these elements, it's good to think about which principles guide you most strongly.

Finally, think about how much time and effort you want to spend maintaining your home or working in your garden, as this will affect the material and design choices for the building and landscaping.

Once you've mapped out your thoughts and principles, you can start scrapbooking images of homes, rooms and gardens that reflect your needs and style — whether for overall aesthetic and palette, particular design features, or examples of homes by architects or designers who you might want to work with.

Putting together textural mood boards can help give your architect or designer an understanding of how you'd like your home to feel. Photos: Lisa Day

The Seatoun House in Wellington by First Light Studio is a small home with a big presence. At just 84 sqm, it was a 2017 NZIA Wellington Architecture award winner in the Small Homes category. Photo: Jason Mann

ARCHITECT'S TIP: TWO CORE IDEAS

Architect Lisa Day of Donnell & Day recommends choosing two core ideas as the ultimate guiding lights for your project — the things that are absolutely most important to you. That way, when you find yourself faced with tough choices, you can go back to your principles and use them to guide you. Some examples of core ideas might be:

- a place of calmness
- durable and fun
- a healthy home that connects our family
- preserving the heritage of the house
- light, bright and beautiful
- on time and within budget
- a close connection with nature.

BUDGET

Whether your budget is tight or generous, you can be assured that it's possible to build a modern eco home within it. Working out your budget is not a straightforward process, and it may undergo several revisions under the guidance of your architect, designer, quantity surveyor, renovation specialist or builder.

Building an eco home, or simply a home with a higher standard of insulation, airtightness and design, will generally cost 10–15 per cent more than the cost of a similarly sized standard home — but this is an investment that will pay itself back over time (usually within a decade), while giving you a more pleasant and comfortable home for all those years. Having said that, cost per square metre isn't a great way to think about your home. An efficiently sized, well-designed home with high-quality fittings will have a higher cost per metre than a large, standard-quality, off-the-plan home, but the former will be a much better place to live in. There are ways of making smaller spaces feel as generous as large spaces — and lots of examples in this book of very livable compact homes.

Remember that your build budget will exclude GST, consents and legal fees and surveying costs. Call your local council to get estimates of these fees and costs for your area. If you fill out and submit the consents yourself, you can save a fair bit of money, but be warned — it can be a time-consuming and frustrating process.

And remember to set aside 20 per cent of your build budget at the start for 'contingencies'. This is for everything that ends up costing more than you expected, that you forgot to add to the initial budget, or to fix things that go wrong.

KEY TEAM MEMBERS

ARCHITECT, ARCHITECTURAL DESIGNER OR DRAUGHTSPERSON

If you have an extremely straightforward renovation or extension plan in mind, an architectural draughtsperson may be all you need to get your ideas formalised on paper. But for new builds or transformative renovations, an architectural designer or registered architect will bring a wealth of knowledge in spatial design, creative thinking and project experience that makes the extra investment worthwhile. This person will be your key partner for your project.

Registered architects and architectural designers are not the same thing. A registered architect has a five-year double-degree qualification, at least three to four years of work experience, and has passed a registration exam following the work experience. An architectural designer may have a two-year qualification or have completed more than half of an architectural degree, but does not have formal registration. A draughtsperson will be a Licensed Building Practitioner (LBP) with draughting skills, but may not have a qualification. A misperception about architects is that they're only for people who are building large, luxurious homes. There are excellent architects who work with budgets right across the spectrum. Don't write off the possibility of having a bespoke home just because your budget is tight.

Take your time and meet a few architects or designers before choosing one. Ask around friends and family who have built or renovated for recommendations, or search online for 'green', 'eco' or 'sustainable' architects in your area. Most local home and living magazines have websites where you can search for features about sustainable design or eco homes.

The New Zealand Institute of Architects has an online tool for finding architects in your area, and the New Zealand Green Building Council lists Homestar practitioners by location. Many architectural firms have websites and social media accounts where you can browse through examples of their work and read their 'approach' or 'philosophy' statements to learn whether sustainability is important to them.

Some questions for potential architects or designers:

- What does sustainable design mean to you, and how do you normally incorporate it into your work?
- If using certain sustainable materials is important to me, do you feel confident incorporating them in your design?
- Do you design to Homestar ratings standards? If not, what sort of energy and thermal performance standards do you work to?
- How many high-performing sustainable homes have you designed?
- Do you do post-project reviews with your clients? Do you have references from previous clients available?

BUILDERS AND BUILDING COMPANIES

Architects and builders work closely together, so your architect may be able to recommend builders who they like to work with. You can also specify your own building company. Either your builder or your architect can work as the project manager for your build or renovation, or if you want to save money and feel confident in your expertise, you can project-manage it yourself.

Your builder is responsible for translating your architect's plans into reality, and will usually be responsible for sourcing the products to do so.

When they look at the plans and product specifications, make sure they

WORDS TO BE WARY OF

If the designer, builder or company you're talking to describes their designs as 'meeting the legal standard', think twice — another way you could put that is 'the worst house we can legally build'. Legal minimums in the Building Code are well below best practice. Insist on higher standards for insulation and windows as a starting point.

know which ones have been specified for their eco-credentials and aren't to be substituted. Sometimes builders will substitute cheaper similar products from the ones specified in a well-meaning attempt to save money.

As with architects, asking friends and family for advice is a good place to start. The NZGBC also lists builders who have delivered Homestar-rated homes. As standalone builders (rather than design-and-build companies) tend not to promote themselves on the basis of a sustainable building approach, what may be more important than their experience building eco homes is their willingness to embrace the challenge.

Some questions for potential builders:

- Are you registered with a professional industry association?
- Do you get multiple quotes from subcontractors?
- How will you manage construction waste on site?
- What is your experience in sourcing sustainable products and alternatives, such as salvaged or eco-certified materials?
- Will using salvaged materials add to the labour costs?
- Will you supply full maintenance information for the house?

DESIGN-AND-BUILD FIRMS/GROUP HOME BUILDERS

These companies offer a complete package of design and construction, either by picking one of a set of pre-designed homes (that can be tweaked to a certain degree) or by working with you to create a semi-customised design, often using a modular system. Land may or may not be included in the package. Since the company has to work within a confirmed price and timeframe, they will understandably want to minimise changes and risks, leaving you with a lot less flexibility.

INTERPRETING ARCHITECT-SPEAK

Architects are generally terrifically clever people, and they work in a complex specialist field. After a while, they become so familiar with the language of architecture that they can forget that civilians don't speak it. This is a quick guide to some architectural terms that may otherwise leave you bamboozled.

- Articulation: The way spaces are joined together. Spaces can be highly articulated (joined together but still very distinct from each other, like in 'broken-plan' living) or continuous (not very distinct from each other, like in 'open-plan' living).
- Circulation: The areas where people move through a space or building. A hallway is 'dedicated circulation', meaning a space designed just for people to move through.
- Fenestration: Windows — from the Latin word for 'opening for light', *fenestra*. (A related word, defenestrate, means to throw someone out a window, which you or your architect may feel like doing at some point of your project.)
- Massing: The shape, 3D form and size of your home, including its sense of verticality or horizontality, lightness or heaviness, and the play-off of solid forms and voids (empty or 'cut-out' spaces).
- Parti: The 'big idea' or concept behind your home's design. It may be presented as a simple statement or diagram.
- Programme: The list of rooms you want in your home. A programme for a typical family home might be three bedrooms, two bathrooms (including one ensuite), an open-plan living/dining/kitchen, a study/media room, laundry and two-car garaging.

There are a number of companies in New Zealand advertising themselves as 'green' or 'environmental' home builders. Make sure you're very clear on what makes their homes 'green' compared to others, and how they measure it. Find out which 'eco-features' are standard, and which are upgrades.

Another consideration when working with a group home builder or developer is that you are putting all your eggs in one basket. If the company goes under, you may be left considerably out of pocket. Make sure that the company has plenty of experience in building the kind of home you want (if they are a new company, with fewer than 10 built projects, they may not have worked out all the kinks in their system), and is in good financial health.

Some questions for potential building companies (you can also ask them relevant questions from the architects' and builders' lists):

- How long have you been in business, and how is the financial position of the company?
- How many homes have you built?
- Have you built on a site like ours before?
- When will I be able to move in? Is there a contracted payment penalty if we are not in by that date?
- Will I have a single contact from the company throughout the build?
- How do you measure the performance standards of your homes?

'Green' builders can be hard to find, so focus on finding someone with an open mind, who pays attention to detail.

This home in Auckland was built using a modular post-and-beam system by Box Living, an eco-oriented design-and-build company. Photo: Emma-Jane Hetherington

ECO-DESIGN AND HOME PERFORMANCE ADVISORS

If you're renovating your home to make it more sustainable and healthy rather than building new, you may not have a clear idea of how to assess or prioritise what needs to be done. So, for the same reason that you wouldn't rely on your own skills to check your car for its warrant of fitness, it's a good idea to get the advice of an independent expert on how your home could be made more healthy and eco-friendly.

The Eco Design Advisors service is a national collective of consultancies that work with eight councils throughout the country to provide free two-hour consultations. It grew out of BRANZ. If you are starting out on the path to making your home more sustainable, booking a session with one of their advisors is an excellent decision. They will go through your home looking at aspects such as your heating, insulation, possible sources of dampness and draughts, and windows.

Beacon Pathway, a healthy homes research and advice service, also offers a contact list on their website (www.beaconpathway.co.nz) of certified home performance advisors who can give unbiased advice on home sustainability and health. Some of the people on the list provide the service for free, and others have a cost attached to their services. As well as pointing you towards the bigger structural issues you might want to address, an eco-design or home performance advisor can give you a number of tips for easy lifestyle changes or quick fixes to improve your home's efficiency right away.

CONSENTS AND PERMITS

All building or substantial renovation work on new or existing homes needs to meet the requirements of the New Zealand Building Code, and anything that affects the building foundations, structure, insulation or the weathertightness of the building will need a building consent to proceed (and will need to be built by a LBP). You may also need resource consents for the way you use or subdivide land, use water, discharge contaminants or affect your neighbours, sightlines or the local environment. Failing to get the right consents may result in hefty fines or infringement fees.

It's a good idea to get your design as close to finished and fully costed (either by the builder or a quantity surveyor) as possible before applying for consents, because if you have to make design changes later on, you may have to resubmit the consents.

Some of the more progressive councils in New Zealand offer to waive or reduce fees for sustainable projects, such as installing solar water-heating panels, hot-water heat pumps, wood-burner wetbacks, rainwater tanks, double-glazing retrofits and solar-power systems. Check with your local council if they offer these (and give them a hard time if they don't).

Kiwibank also offers a 'Sustainable Energy Loan', where they'll give you a contribution towards sustainable energy systems and waive certain loan fees if you're an existing customer.

DISPUTES

The best way to avoid disputes is to have a good contract in the first place, and make sure that all parties are absolutely clear on what it includes. If your residential project costs $30,000 (including GST) or above, by law you need to have a contract, but it's a good idea to have one in place for smaller jobs, too. The Ministry for Business, Innovation and Employment has excellent advice on what should be in or understood around a building contract, including the process for resolving disputes, at www.building.govt.nz.

CHAPTER 5

FOUNDATIONS AND FLOORS

Rammed earth, polished concrete and shiplapped vertical cedar make a beautiful interplay of colours and textures at the entrance of the Rammed Earth House by Assembly Architects in Wanaka. Photo: Simon Devitt

Above: The K Valley House by Herbst Architects incorporates both concrete-slab foundations where the house cuts into the hillside, and a suspended timber floor to allow the living areas to 'float' above the landscape. Photo: Patrick Reynolds

Opposite: This home, designed by Makers of Architecture in collaboration with the client, a Canterbury based artist, is a mere 72 sqm. The inverted roof truss divides the ceiling into two planes, one running flat over the kitchen/utilities and shared circulation space, while the other rises upwards, expanding the feeling of spaciousness in the living and bedroom areas. Photo: Makers of Architecture

In the years since 1970, the size of the average New Zealand home has gone from just under 130 sqm to a little over 200 sqm. That's nearly a 60 per cent increase in the footprint of each new home, despite the fact that buildable land is becoming increasingly squeezed in our main cities.

What is the extra space being used for? In large part, more stuff. More parking spaces for more cars: in the 2013 Census, almost 55 per cent of households were registered as having two or more vehicles, with the number of households with one or no cars declining steadily since 1991. Additional bathrooms: most new homes have an ensuite to the master bedroom, and sometimes to secondary bedrooms as well. Media rooms where we can park an extra-large-screen television in front of an expansive couch system, bigger kitchens for our bigger fridges and more closet space to fit the increasing amount of inexpensive fast fashion that we are buying. In other words, much of the space is being used to accommodate our addiction to consumer spending, and is adding weight to our carbon footprint through encouraging greater use of power, water and fossil fuels.

With an increasingly multicultural population, we are also seeing more families where it is traditional for grandparents to reside with their children and grandchildren. That means more bedrooms, more bathrooms and more living spaces. Larger isn't necessarily bad, if the space is being used well and the house is airtight and well insulated to make it energy-efficient — even more so if the home generates some or all of its own power. In the case of the multi-generational family home, having a larger house can mean creating less of a load on the environment than having a separate dwelling for the grandparents.

But when the extra square meterage is being added because the owner has an unrealistic idea of how often the spaces will be used — like adding bedrooms for adult children to come and stay in, even though they might sit unused for most of the year, or having a separate dining room that is only used on special occasions — the additional space adds a huge and unnecessary cost in both financial terms and on the environment. During the research for this book, many architects stated that clients often came to them with an extravagant idea of the house size they wanted, only to be convinced that downsizing the footprint would allow them to invest in better-quality materials, fittings and fixtures, and make a home that was more comfortable with no genuine loss of usable space. NZGBC CEO Andrew Eagles says that shaving as little as 4 sqm off an average build size can pay for an upgrade on the remaining square footage that will result in a house that 'works much better'.

Several of the homes featured in this book have compact footprints, yet they don't feel small because the architects have designed them cleverly to give a feeling of spaciousness — lifting ceilings, doubling up on space use and letting in plenty of light.

There are two main types of residential foundation construction in New Zealand: concrete slab or suspended floors (usually timber, but sometimes cross-laminated timber — CLT — or concrete) over a crawl space or basement. The majority of new homes are built on a poured concrete slab. The type of foundation you choose will depend on the geotechnical requirements of the site you have and how you want your building to perform.

CONCRETE SLAB

At a basic level, concrete is not what you could call a 'green' building material. Although it's made from naturally derived materials (sand, gravel, water and cement, which in turn is made from limestone, clay shale and gypsum), the extraction of the materials is extremely energy-intensive, and the process for making cement releases huge amounts of carbon dioxide. Despite sand being perceived as a near-infinite resource, demand for concrete sand is a growing issue, according to environmental construction website BuildingGreen. Desert sand, scoured by the wind, is too smooth to be used for concrete production, so construction sand has to be dredged from lakes, riverbeds, beaches and the ocean floor, causing damage to marine ecosystems.

Concrete production accounts for around 5 per cent of the world's annual carbon dioxide emissions. Close to 90 per cent of the carbon emissions produced for every tonne of concrete comes from the cement-making process, even though cement makes up only around 14 per cent of a standard concrete mix.

Cement is made by quarrying limestone and clay shale, which are then crushed and ground together. The resulting mixture is burned in a kiln at temperatures between 1400 and 1600 degrees Celsius. As water is released from the mixture and the material decomposes, carbon dioxide is released and calcium silicates form. At the end of the process, marble-sized balls called clinker are left. These balls are ground to a powder and mixed with gypsum to make what's called Portland cement (named for its similarity to Portland stone, from Dorset, England).

Basic concrete is made up of aggregates, both fine (sand) and coarse (usually gravel and crushed stone), mixed with cement to act as a glue or binder. When Portland cement is hydrated by having water added to make it into a paste, it slowly hardens into the familiar solid, durable mass that we see every day in driveways, footpaths, buildings and walls.

As well as sand and gravel, concrete companies have been experimenting with alternative aggregates to make concrete more eco-friendly. Demolition waste, crushed bricks, shells, asphalt, crushed glass and even shredded car tyres can replace a percentage of aggregate in a concrete mix. Other substances called pozzolans can also be added. Pozzolans are waste products that act in a similar way to cement, so they decrease the amount of cement required, making the resulting concrete a little more eco-friendly. Two common pozzolans are fly ash (a combustion product from burning coal) and slag (a by-product of iron smelting). New Zealand concrete company Firth offers a 'sustainable' EnviroBlend concrete that carries Environmental Choice label approval.

Steel reinforcing is often used to improve the tensile strength of concrete. Overseas, carbon mesh reinforcement is being explored as a more environmentally friendly alternative to steel, as it weighs four times less and offers five times the tensile strength.

Despite the embodied energy and resource consumption of concrete, it has two qualities that allow it to be considered for use in an eco home. First, it's durable, lasting for decades without the need for substantial maintenance. Second, and more importantly, it's an ideal material for thermal mass, particularly as a floor slab.

Concrete stairs leading up to the entrance of the Wright House by Assembly Architects in Arrowtown. Photo: Simon Devitt

WHAT IS THERMAL MASS?

Thermal mass is the ability of a material to store heat from sunlight falling on it, and slowly release it later when the sun moves away. This stabilises temperatures in a room in two ways. First, when sunlight shines directly on thermal mass, the mass absorbs the heat. Even when the direct sunlight has moved away from the mass, if the ambient temperature is higher than the temperature of the thermal mass, the mass will continue to absorb heat, keeping the room at a more stable temperature and reducing overheating. Second, in the evening, when the room starts to cool, the thermal mass begins to release stored heat, again stabilising the temperature. Dense materials like stone, concrete and rammed earth have a high thermal mass, while timber has a lower thermal mass (although densely constructed timber products like cross-laminated timber do have a certain amount of thermal mass). To be effective as thermal mass, a concrete slab needs to be 100–200 mm thick, and be directly exposed to sunlight or covered with a dense material like tile or stone. A carpeted concrete-slab floor won't give the same benefits.

Of course, it's not as simple as banging down some concrete and sticking a big window next to it. Thermal mass needs to be incorporated as part of a careful passive-design strategy that includes building orientation, appropriately sized windows that face the right way, shading during peak sunlight, ventilation and insulation. Thermal mass that gets too much sun exposure can result in the house overheating, especially during the summer, making it uncomfortable to live in. And a concrete floor slab that's poorly insulated will lose much of its stored heat to the ground or exterior.

SLAB INSULATION AND THERMAL BREAKING

Thermal breaking is an insulation technique that means separating adjoining materials that conduct heat easily by inserting another material that does not conduct heat well in-between in order to slow the loss of heat. It's often used in relation to window frames and walls, but applies to slab floors as well.

A basic concrete-slab foundation is simply a concrete pad poured on top of a layer of gravel and damp-proofing membrane, with a deeper reinforced concrete footing around the perimeter. Under the Building Code, a concrete slab with an R-value of 1.3 (see sidebar on page 100) does not need to be insulated, but it can be — and should be, given that according to BRANZ, 80 per cent of heat lost from a concrete slab is through the edge. (Slabs don't lose much heat through the ground underneath because ground temperatures are much more stable than air temperatures.) If you have a hydronic underfloor heating system (see Chapter 9: The Living Spaces), you definitely want to insulate or thermally break the slab, or you'll be paying for heat to flow straight to the outside of the house.

There are a number of systems available for slab insulation and thermal breaking. Polystyrene insulation can be laid underneath the floor slab, a

Afternoon sun pours onto the concrete slab floor of the 339 House by Strachan Architects in Mt Eden. Photo: Simon Devitt

thermal break can be created between the edge of the concrete slab and the perimeter footing, or the exterior of the perimeter footing can be insulated. New Zealand company MAXRaft offers a concrete-slab foundation system that is fully insulated, including the exterior, and has an R-value of 4.5 — almost three times as warm as the Building Code standard, and costing only about 10 per cent more than an uninsulated concrete slab.

SUSPENDED FLOORS

Suspended floors are more frequently used on sloped or uneven sites, and for second storeys, and are found in almost all homes built prior to the 1970s, or about 80 per cent of New Zealand homes. They're usually constructed from timber joists suspended over concrete perimeter foundation walls or piles, sometimes with a layer of sub-flooring, and then finished flooring above. Cross-laminated timber panels can also be used as a standalone suspended flooring surface. Suspended concrete floors built on steel joists, traditionally used in commercial building and even carparks, are becoming more popular in residential construction, and allow second-storey rooms to have polished concrete floors.

INSULATION AND RISING DAMP

Many older homes with suspended timber floors have minimal or no underfloor insulation. Uninsulated floors can cause up to 14 per cent of the heat loss from a home, so if your home doesn't have underfloor insulation, it's worth putting it in. Some houses will have reflective-foil underfloor insulation, which can be a safety hazard due to the risk of electrocution when working on it. Never touch it unless the power to the house has been turned off. Foil insulation cannot be repaired, and if it's damaged should be replaced. Bulk insulation like fibreglass, polyester or wool batts or polystyrene blocks are generally your best bet, and polystyrene blocks may be the easiest to install in low crawl spaces under bungalows and villas. Energywise advises that you choose underfloor insulation with an R-value of at least 1.4, as the standard R-value for floors in new homes is 1.3.

A related issue is the presence of damp ground underneath suspended timber floors. Damp ground can release up to 40 litres of moisture every day, according to research by BRANZ. That moisture can compromise the performance of your underfloor insulation and even rise into your home, damaging your carpets and rugs, and causing 'rising damp' mould on your walls near the floor. If the ground in the crawl space or basement underneath your house is not dusty and dry, check that the air vents in the walls of the space are clear of dirt and obstructions so that air can flow through.

An inexpensive way to significantly reduce the problem of rising damp is to install a polyethylene sheet underneath the house. This can be purchased along with tape from a hardware store for a couple of hundred dollars. The sheet should be laid out flat directly on the ground throughout the entire underfloor space, lapping up around edges, corners and piles, and be taped securely in place. Bricks and stones can be used to weight it down in case wind gets underneath it. The Eco Design Advisor website (www.ecodesignadvisor.org.nz) has a good instructional fact sheet on how to lay a moisture barrier. Reducing moisture rising into your house won't make it warmer in itself, but it will make it easier to heat.

FLOOR COVERINGS

While the foundations are the first part of a home to be laid, the flooring that goes directly over the top is often the last part to be finished (because you don't want to lay your new flooring surface before you've plastered or painted your walls), making a nice conceptual bracket for your project.

Aesthetic preferences and budget aside, the key qualities you'll be looking for in an eco-friendly flooring surface are durability, inertness (no or low off-gassing), a pleasant tactility, ability to be cleaned easily without harsh chemicals and recyclability at the end of its life. Although some flooring surfaces have an insulation value, this is not hugely important, as minimum R-values for floors are calculated without including floor coverings.

CONCRETE

Polished concrete has become increasingly popular as a flooring finish, despite early perceptions that it was cold and industrial. As people became more aware of its benefits as a source of thermal mass, they also began to enjoy its other qualities — the natural appearance of the aggregate, a contemporary minimal aesthetic, its ease of cleaning and its resistance to damage. It's impervious to moisture, so can be used as a flooring surface in bathrooms, laundries and kitchens, although it needs to be sealed with an oil for protection. This will last about five years before it needs resealing. Concrete can be polished to a range of finishes, from light to darker with more exposed aggregate, and can come in a range of tones, from cool to warm.

Concrete-panel flooring is a relatively new addition to the market, and offers the possibility of retrofitting suspended timber floors with a concrete finish. The concrete in these tongue-and-groove panels is a skim layer over a 14 mm magnesium oxide cement that contains recycled textile, meaning that while they absorb ambient heat, they do not give the same level of thermal mass as a poured concrete floor and are primarily a decorative finish. However, the magnesium oxide and recycled textiles used in the panels means they have a substantially lower embodied energy than poured concrete.

Solid polished concrete floors in the 339 House by Strachan Architects absorb the heat from the sun. Photo: Simon Devitt

TIMBER

Many villas, bungalows, Art Deco and Modernist homes have beautiful existing hardwood floors in native timbers like mātai, rimu, tawa, kahikatea and miro that can be difficult to procure new. If you have one of these floors in your home, it's practically criminal to cover them up with carpet or paint. Many have uneven finishing and sealing, and will need to be sanded and resealed with a zero-VOC finish or natural oil rather than polyurethane, which off-gases VOCs. The work doesn't necessarily need to be done to the quality of a new timber floor. The existing nail holes, wall lines and even staining can be quite beautiful, telling a story about your home's history.

If you don't have an existing timber floor, but love the look of old timber, salvaged native timbers can still be found, but may need extensive preparation before they can be used. Imported salvaged hardwoods like teak and oak are available, just look for ones that have been FSC- or PEFC-certified. New hardwoods should come from certified sustainable plantations that don't replace natural rainforest, but with imported timbers this can sometimes be hard to ascertain. Many flooring companies sell woods that are both certified and uncertified, so don't assume that because you've looked at one certified option, the rest will be, too.

Of course, solid timber floors aren't cheap, and they can be prone to cupping, shrinking and bowing, so there are a number of alternatives designed to give greater stability with the look of timber at a lower cost.

ENGINEERED WOOD

Cross-laminated timber (see page 98) can be used as structural flooring with an exposed, sealed finish that can match walls of the same material. This needs to be specified as part of a new building, and would be difficult to retrofit. It won't shrink or warp, and is extremely structurally stable. An advantage of a cross-laminated timber system is the very clean profile at joins between the floor and wall — no skirting boards needed.

Engineered wood involves a timber veneer applied over two substrates, usually plywood or high-density fibreboard. The timber veneer is thick enough (2.5–4 mm is normal) to have an appearance close in quality to solid wood (which is usually 13 mm or 19 mm thick), and can even be lightly sanded down and refinished. Depending on the thickness of the veneer, the price can start to approach that for solid wood. Look for engineered wood that is FSC-certified and uses plywood with an E0 emissions rating. Emissions ratings describe the amount of VOCs that are released by a building product per litre of air. The exact amounts allowed to qualify for each level can vary depending on the product and country where it was labelled, but in general an E0 or E1 emissions rating is considered acceptable for indoor use.

Laminate floors have a very thin pressed-wood veneer with a printed image that looks like a genuine wood panel, and a resinous seal laminated

Above: At the Lynds House in Queenstown, the Xlam CLT floor structure provides a clear span that does not need additional structural support. Photo: Makers of Architecture

Far left: More than one type of timber can be used effectively in a home. The oak floors in this home by Rafe Maclean contrast effectively with the cedar walls. Photo: Simon Devitt

Left: Matai flooring was hiding beneath carpet at a 1950 home. It was sanded down and refinished with a low-VOC acrylic resin. Photo: Anna Kidman

over the top. The laminate is then laid over the top of a foam underlayment, which is likely to be made from petrochemical products (although may contain recycled content) and may give off VOCs. When inspected closely, it's pretty obvious that it's not real wood, but at standing height it can be harder to tell. Laminate is the lowest-cost option to achieve a similar look to timber. The tiles click together in a tongue-and-groove type system, like real wood.

Engineered wood and laminate are much quicker to install than solid wood, as they are ready-made products, whereas solid wood floors need to be finished on-site.

BAMBOO

Commercial bamboo flooring first came to the local market in the early 2000s. It went through an initial stage of being fêted as the most eco-friendly flooring 'wood' (bamboo is actually a grass) available. Over time, some concerns around lack of certification, the durability of prematurely harvested bamboo, carbon kilometres and the adhesives used in its manufacture resulted in a waning of its popularity. Today, many of the early concerns have been addressed and bamboo is again a viable option for eco-friendly flooring.

The most common type of bamboo is *Phyllostachys edulis* or Moso bamboo, which is largely sourced from the Hunan Province in China. It's very fast-growing, replacing itself in just six years, and sequesters carbon in the same way as trees. Strips of bamboo are glued together to create floorboards, which may use VOC-emitting urea-formaldehyde resins. Lower-emitting alternatives are available, so look for these — credible certifications are Greenguard, E0 or FloorScore. Forest Stewardship Council certification began for bamboo in 2008, and now many manufacturers hold it.

CORK

Popular in the 1970s and 1980s, cork tiles are seeing something of a revival. Older-style cork tiles had an orange cast due to the treatment used to seal them, but newer ones come in a range of much more subtle and natural colours, from almost white, to dark brown and black.

Like rubber, cork is periodically harvested from living trees and does not require the felling of the tree. Some cork tile products also contain a mix of pre-consumer cork recycled from the wine-cork industry. Cork has a resilient softness and acoustic absorption that makes a room quieter and more comfortable, but its soft surface means that it can be susceptible to damage. A dropped glass may bounce when it hits a cork tile, but a dropped knife is likely to leave a hole. Dragging a heavy object across it will likely leave scrape marks.

Thanks to the tiny air pockets in its honeycomb structure, cork also acts as an insulator. It's not suitable for bathroom or laundries, where there is a risk of flooding, but may be suitable in kitchens. Cork can also be used as an underlay for carpet or laminate instead of petrochemical foam. Portugal is

Left: This 1971 home was given a modern renovation by Bureaux using materials that were authentic to the era — such as cork flooring — but presented in a modern way. Photo: Simon Devitt

Right: Salvaged wood rounds set in concrete make a hard-wearing floor surface for the entranceway of the K Valley House by Herbst Architects. Photo: Patrick Reynolds

the largest producer of cork in the world, which means that cork products are likely to have travelled from the other side of the globe to get to your home.

STONE, BRICK, CERAMIC AND TERRACOTTA

Tiled floors represent the ultimate in durability in high-traffic and wet areas. Most tiles are made of extremely durable and naturally variegated or veined materials, which add a tactility and sense of connection to the outdoors — and indeed, some tile flooring can transition seamlessly from indoors to exterior paving. They are fire-proof and spill-proof, and unless you go for a particularly emphatic colour or design, are quite timeless in their appeal.

If the cool, industrial look of concrete isn't to your taste, warmer-coloured tiles, brick and stone are ideal surfaces to lay over underfloor heating systems, as they act as a thermal mass to store and release heat. Manufactured tiles and bricks are baked in kilns at high temperatures, so have a high embodied energy, which is somewhat balanced by their long lifespan. Stone doesn't require firing, but the quarrying process has its own energy costs, as well as the environmental impact of extraction.

Stone: There are a wide variety of stones available for flooring in New Zealand, although most flooring and paving stones are sourced from overseas; Indonesia, India, China, Italy, Spain and Turkey are common supply countries. Sandstone, marble, granite, travertine, slate and limestone are popular choices, with each offering different colours and characteristics. Some salvaged stone is available — even broken or irregular stone tiles can make a beautiful mosaic floor finish when done well.

A key factor to be aware of with stone is the level of moisture absorption — non-vitreous stone tiles absorb moisture and should not be used in wet areas, while vitreous and impervious stones are resistant to moisture and can be used more widely. All stones need to be sealed to some degree, with more porous stones requiring greater levels of sealing. Acidic cleaners can damage this sealing, resulting in a need to reapply.

Brick: Brick flooring is often seen in outdoor spaces in New Zealand, but is less common as an interior flooring finish. It can work well in high-traffic or transition areas like entranceways or living areas with a lot of indoor–outdoor flow. Bricks are a comparatively low-cost choice, available in a range of earthy

Original ceramic tiles can be preserved or recycled in a renovation. Broken or chipped ones can be repurposed in a mosaic form. Photo: Emma-Jane Hetherington

colours, and develop a soft patina over time as they wear down. They tend to retain warmth longer than stone, but conversely, also take longer to warm up. They're made widely in New Zealand, reducing the carbon-kilometre cost of transporting them.

Ceramic and terracotta: Hard-baked ceramic and terracotta tiles are a versatile flooring option, with a vast range of colour and shape options. Glazed ceramic tiles tend to be used in wet areas rather than in living spaces, largely because they're cold to walk on. There's one main exception: porcelain, which is fired at a higher temperature, and is more durable. Wood-look porcelain tiles are a living-area option that has been growing in popularity as a less expensive alternative to real wood, but work best with an underfloor heating system to make them comfortable year-round.

As a hard, smooth surface, tiles are easy to clean and don't harbour allergens like dust, pet hair and pollens. The grouting between the tiles is a different story, however, and can attract mould growth if it's not sealed well and cleaned regularly. Reasonably harsh cleaners or a great deal of elbow grease can be needed to keep pale grouting in wet or spill-prone areas looking good.

CARPET

One of the two most popular choices for interior flooring finishes (alongside wood), particularly in bedrooms and living rooms where comfort is king, carpet is widely available, relatively inexpensive, carries an insulation benefit, and comes in a vast range of colours, textures and even patterns. Its softness means that it's safer for tumbles and falls by small children or elderly homeowners, and it absorbs noise, making rooms quieter.

On the other hand, carpet traps dust and allergens (although some studies have suggested that rather than aggravating allergies, carpets decrease the amount of particulates in the air, making the indoor environment healthier), can be hard to remove spills and stains from, and needs to be replaced more frequently than hard flooring.

It's this frequent replacement of carpet that is perhaps its greatest environmental impact. Most old carpet and underlays are sent straight to landfill, because they are made of complex, often petroleum-based synthetic-mix fabrics that are difficult to recycle. More than 2 billion kilograms of old carpet are sent to landfill in the United States every year (comparable New Zealand figures are not available).

So the big questions to ask yourself when buying carpet are how long the carpet is going to last, and what you (or the next owner of the house) are going to do with it when it comes to the end of its life. If you're laying carpet in a high-traffic area, it's best to invest in the most durable one possible, and take good care of it so it lasts longer. If the carpet still has some use in it at the time you're disposing of it, it may be able to be recycled or reused. In

Seagrass rugs bring a natural warmth and texture to a living space. Photo: Elizabeth Goodall

Auckland, Premium Flooring runs a recycling programme, in Christchurch, EcoDrop accepts carpets in good condition, and Cavalier Bremworth recycles wool carpets for their Flashbac recycled carpet underlay through the carpet recycling bins at Envirowaste centres.

Natural fibres: Wool carpet is the original carpeting fabric, with all the eco-qualities that you expect from a locally sourced renewable natural material — breathability (absorbs and releases moisture), thermal insulation, low embodied energy, zero-VOC and biodegradability. However, it is more expensive than nylon, fades faster and can be harder to remove stains from. Wool carpets usually have a hessian or jute backing, but if they are attached to a synthetic backing, this may emit VOCs.

Sisal carpets are an expensive but hard-wearing and eco-friendly option with a distinctive woven appearance. Sisal fibre comes from the agave plant, which is mainly grown in Asia, South America and Africa, so it has a higher carbon-kilometre count than locally sourced wool. Sisal fibre is also coarser than wool, so wool carpets or synthetic woven to resemble sisal are sometimes preferred. Jute and seagrass are other natural fibres that are more commonly used as rugs and mats than carpeting.

Tatami mats, made from woven rushes, have been widely used as flooring in Japan for centuries, and many modern Japanese homes still have at least one 'tatami room'. Modern tatami mats usually have a polystyrene or particleboard core that can emit VOCs, so although tatami are renowned for the scent released by the dried woven rushes that cover them, make sure to enquire about the emissions rating of the core.

Synthetic fibres: Solution-dyed nylon is the main synthetic fibre used in carpets, though there is also polyester, polypropylene (also known as olefin) and triexta (a new product that uses corn glucose to replace some of the petroleum in the production process).

Synthetic carpet fibres are made from petroleum products, do not biodegrade and are resistant to fading, although they do off-gas VOCs. Some nylon carpets, such as Cavalier Bremworth's Siren range, are made from 100 per cent recycled nylon fibres, making them a more eco-friendly choice. These carpets may still be difficult to recycle at the end of use, however. In early 2017, the US carpet company Mohawk launched the Airo range of fully recyclable carpets — the first in the world — which are made from a single material, polyester.

Carpet underlay: Synthetic foam underlays can off-gas VOCs and contain brominated flame retardants that have been flagged by environmental watchdogs as a health concern. Look for underlays that are certified as low-VOC, contain recycled content or are made of natural, non-off-gassing materials like cork.

Linoleum and vinyl: The term 'resilient flooring' covers a range of continuous or sheet flooring materials that tend to be more accessibly priced, though they can be prone to fading and damage from dragged or heavy furniture and sharp objects.

Linoleum, which has been around since the late 1800s, is a naturally derived material made from linseed oil, ground cork or sawdust, pine resins and calcium carbonate on a canvas or burlap backing. It was hugely popular as a floor covering until the 1950s, and was still frequently used as a kitchen, laundry or bathroom flooring material for the next 30 years. Today it tends to be used more in commercial spaces.

Sheet vinyl has become the main replacement material for linoleum, and the words are sometimes used interchangeably. PVC (vinyl) sheeting is a considerably different product, however (see sidebar on page 88). It's low-cost, waterproof, durable and comfortable to walk on, being warmer and having more 'give' than unheated tiles. Vinyl needs to be laid over a smooth substrate or any bumps and dents underneath will eventually show through. It can be damaged by furniture and appliance drag or weight, and if ripped needs to be fully replaced. Vinyl tiles, which are often printed to resemble wood, are a solution to this, as single tiles can be replaced. Vinyl does release VOCs, so look for ones with low-emissions certifications.

IS PVC SAFE?

Polyvinyl chloride — also known as PVC or just vinyl — is a widely used and versatile plastic. In harder forms, it is typically found as pipes and spouting, door and window framing, plastic containers and exterior siding. Adding phthalate plasticisers to PVC makes it more flexible, so it can be turned into hoses, flooring, wallpapers, wiring coatings, blow-up toys, records, swimming-pool liners, shower curtains and those terrible, shiny pants and skirts that were briefly popular in the 1980s and 1990s.

PVC is made from the organic compound vinyl chloride (vinyl chloride monomer), which is manufactured from the petroleum product ethylene and chloride through the heat-, refrigerant- and water-intensive process of chlorination and cracking, which creates by-products of hydrochloric acid and dioxins.

In the 1970s, vinyl chloride monomer exposure was linked to a rare form of liver cancer in vinyl factory workers, leading to lower exposure standards being set to protect worker health. Some environmental and health groups also express concerns over additives in some PVCs, like the toxic heavy metals lead and cadmium, which can leach into the environment, and the phthalate plasticisers used to soften PVC, which have been associated with endocrine (hormone) disruption. Vinyl also releases VOCs, which negatively affect indoor air quality.

From a health perspective, although the chemicals in PVC production have been identified as health hazards, there haven't been direct studies proving that modern PVC household products have specific harmful effects. Some people prefer to take a 'better safe than sorry' approach to PVC products, especially ones that come into contact with children.

The greater concern with PVC may be the environmental effects of the production and disposal processes. When PVC is burned in sub-optimal conditions (such as backyard burning, landfill fires, house fires, hospital waste disposal and cable-stripping), it releases dioxins, which have been identified as a persistent organic pollutant (POP) associated with immune disorders, birth defects and cancer. Needless to say, never deliberately burn PVC.

Although PVC can be recycled (it's identified by the recycling number 3), it gradually downgrades, and eventually ends up in landfill, often before the end of its recyclable life. From an eco-perspective, it's a plastic with a big environmental footprint. Although it's the industry standard for some products (like piping), cleaner substitutes exist for many of its household uses.

CHAPTER 6

THE STRUCTURE

The inside of your walls might be the least sexy and most easily forgotten part of your home, but when it comes to making an eco home, it's hands down the most important part. Investing in good bones for your home will pay off for years to come, both in financial terms and in greater levels of comfort and health.

There are many different approaches to wall-building, and a wide range of materials can be used. Multiple structural techniques and materials can be combined in new builds and renovations with the right design and engineering guidance. Depending on where in the country you live, different aspects of your structure take on different levels of importance. In the south, where temperatures are more extreme, good insulation to keep homes warm in winter and cool in summer is usually the highest priority. In the humid north, excellent ventilation is needed to avoid mould. But that's not to say that the other aspects can be neglected — it's still cold enough in the northern winter to need good insulation for comfort and energy efficiency, and southern homes must still be well ventilated.

A good architect or designer will balance these priorities carefully, choosing materials that perform best in your local climate, and taking into account the embodied energy of the key materials.

FRAMING SYSTEMS

WOOD

Sustainably grown wood is a highly eco-friendly material, thanks to its renewability and carbon-sequestration qualities. As structural framing, it's accessibly priced, lightweight and easy to install, and it performs well under earthquake stress. New Zealand's rich timber resources mean that the majority of our homes have been built using timber internal framing, with insulation batts fitted into the framework. This is colloquially known as 'stick and fluff' building. The minuses are that building with a timber frame can be slower than other types of construction, and the walls have to be sealed and insulated on-site, where they can potentially become damp or dirty.

Villas and bungalows from the late 1800s and early 1900s may have been built using kauri, rimu, mātai or imported Oregon timber, but since the 1950s onwards, locally grown radiata pine has been the most commonly used framing timber. As became painfully evident from the leaky homes crisis in the late 1990s and early 2000s, untreated timber can rot if it becomes damp and does not dry out, so the New Zealand Building Code specifies that most timbers need to be treated with a chemical solution to increase their durability and rot- and pest-resistance. Timber is also often treated with the salt borax to increase its fire resistance.

Until the 1940s, creosote (derived from coal tar) was mainly used to treat structural timber, and older homes may contain this. Today's most common

In 2013, the HIVE High Performance House designed by Salmond Architecture was launched at the Christchurch Home Innovation Village. It featured a new steel framing system called Warmframe, designed to work in conjunction with a high-performance insulation system to give a Homestar 8 rating. It was prefabricated off-site and finished on-site in less than 10 weeks. As a test case for creating warm, dry homes using steel framing it was a success — in fact, the pellet burner used to warm the show home was found to make it too warm even on the lowest setting, so even less heating would be needed. Image: Salmond Architecture

preservative is chromated copper arsenate, or CCA, which is pressure-treated into the wood. The copper helps prevent decay, while arsenic acts as an insecticide, and chromium helps fix the other chemicals in the timber.

Both creosote and CCA are on the Declare Red List of chemicals to be avoided if possible. Arsenic and creosote are both known carcinogens, and, although there is little chance of exposure from treated timber encased in a wall system, there are concerns that incorrectly used or disposed of treated timber can leach arsenic into the ground, potentially getting into waterways. Improper site-handling practices can also directly expose construction workers to these chemicals.

For those who don't want to use treated timber in their house framing, the New Zealand Building Code allows for the use of untreated Douglas fir (which is naturally more durable than untreated radiata pine, though not as durable as treated timber), in low-risk residential buildings that meet certain criteria.

STEEL

Lightweight steel framing came to the New Zealand market in the 1970s, and is a minority choice for use in residential framing in New Zealand — around 5 per cent of the market — although it is growing in popularity. As a material, it's very strong and even more lightweight than timber, fire- and rot-resistant and it doesn't twist or warp when moist, as wood can. Its strength lends itself towards designing open-plan structures with lots of glass, which fits with the Kiwi propensity for indoor–outdoor flow. Overall, the cost of steel framing and timber framing is generally comparable.

Steel's main drawback when it comes to eco-performance is the embodied energy in producing it. Although less steel framing is needed for a house or extension, as a material its embodied energy is dozens of times higher than timber — it is the most energy-intensive structural framing material by weight. You can source steel that has some recycled content, which reduces the embodied energy cost. The higher the proportion of recycled content, the lower the embodied energy cost will be. Choosing New Zealand-manufactured steel (which is produced using iron extracted from West Coast iron-sand beaches) is a better choice than imported steel, as it means accumulating fewer carbon kilometres.

A second drawback is that steel conducts heat very easily, so steel-framed homes must be well designed in order to avoid heat loss through thermal bridging. On the flip side, steel's solidity as a material means it is fully recyclable, and has minimal on-site wastage, as steel frame sections are often manufactured off-site.

STRUCTURAL SYSTEMS

MASONRY: STONE, BRICK AND CONCRETE

Masonry building involves the use of stone, bricks or concrete as solid load-bearing structures that may also work as external and internal finishes. Stone, brick and concrete are building materials for several lifetimes; they're fire- and rot-resistant, and extremely durable (although must be reinforced correctly in New Zealand's earthquake-prone conditions). But the main advantage of solid masonry construction is its high thermal mass, which can be used to capture and store free heat from the sun in winter — and conversely, prevent interiors from becoming overheated in summer.

Although New Zealand has beautiful stone resources, the high cost of building in stone compared to timber means it's a minority choice as a structural material, and is usually used for feature walls, façades or detailing instead. Solid brick homes are very rare, although brick veneer is one of the most popular cladding choices.

Stone, brick and concrete have much higher embodied energy than timber, due to the effort it takes to extract and process them, even when the stone is sourced from close to the building site. Stone is also a non-renewable resource, although under certain circumstances it can be recycled. Locally quarried building stone includes basalt, Coromandel granite (actually diorite), schist, Oamaru limestone, Dunedin bluestone, marble, ignimbrite and scoria. Our most abundant stone is greywacke, but it's a poor choice for building because it doesn't split well.

Cast in-situ concrete walls were used for this Auckland eco home for their 'honesty' and thermal mass qualities, says architect Tim Dorrington. Photos: Emma-Jane Hetherington

Solid concrete construction has a surprisingly long tradition in New Zealand, with evidence of concrete homes being built as early as the 1850s. As a material it was relatively cheap, durable and — once reinforced — solid, and its ability to be made using a range of ingredients appealed to early New Zealanders' DIY ethos.

Hollow concrete masonry units — often called breeze blocks, cinder blocks (when ash is incorporated as part of the mixture) or just plain concrete blocks, were first made in New Zealand in the early 1900s in Wellington by a Dane called Niels Nielsen, who promoted them for use in building 'substantial, durable and cheap' houses. In the 1950s, Sir Miles Warren, a then young architect inspired by Scandinavian design, was one of the first architects to use concrete blocks as a modern, heavier alternative to lightweight timber in the cool Christchurch climate.

Although at first the minimalist, even brutal look of concrete met with bemusement, it soon caught on, and some of New Zealand's most iconic mid- and late-twentieth-century homes feature concrete construction, including Ivan Juriss's 1960 Mann House in Auckland, John Scott's 1970 Martin House in the Hawke's Bay, the 1973 pre-cast concrete Sargent House in Remuera by Fairhead, Sang and Carnachan, and many of Warren and Mahoney's projects. Today, concrete block, pre-cast concrete-slab construction and poured-on-site 'tilt-slab' are popular modern forms of construction.

Above: Some sites, particularly rural ones, contain a natural bounty of stone suitable for building. When architect Daniel Marshall designed a rural home for a Christchurch client in 2015, a stone-block exterior wall was part of the design. But when the foundation excavators discovered plentiful natural stone on the site, he decided to extend the use of stone within the home to take advantage of the free natural resource. Today, it's one of the most dramatic features of the design. Photo: Ernie Shackles

Left: The Rammed Earth House in Wanaka shows how a traditional, eco-friendly building method can result in a beautifully modern home. The earth for the build was sourced locally (though not on-site, as it's important to get exactly the right soil mix for building). Photo: Simon Devitt

EARTH-BUILDING

Earth-building, also sometimes called 'natural building', is the oldest form of building. It's still used extensively in domestic architecture around the world, particularly in rural areas and developing nations, but in developed nations tends to be regarded as a very 'alternative' method of building, and can be slow or difficult to work through.

Earth is universally available as a building material, is low-cost (in terms of material, although labour costs are high) and has little embodied energy. A key attraction for many people interested in earth-building is that they can be part of the building process.

When correctly constructed, earthen homes can last for hundreds of years. Generally, earthen building is better suited to warmer climates, though when paired with good insulation can work in more temperate zones. Earthen homes characteristically have thick walls that act as thermal mass. Because of the wide colour variation in earth, the walls can have beautiful planned or natural gradations in colour. They also have a textural softness, and absorb sound extremely well.

Rammed-earth homes blend earth mixed with cement or another binder and water, which is then pressed into form on-site. This gives a clean, modern look. Adobe (also known as mud brick) building uses bricks made of earth, sand, straw and binders that are often finished with an earthen plaster. Rounded and curved edges and surfaces are common.

This New Plymouth home was built from hempcrete. An internal window in a wall shows the unfinished hempcrete heart of the structure. Photo: Jane Dove Funeau

STRAW BALE AND HEMP

Straw bales — a building material that could hardly be more vernacular given New Zealand's agricultural history — have extremely high insulation value, and are seeing a small but growing return to popularity in modern construction in America, where they were first developed. Straw bales can work as a load-bearing structure, but many natural builders combine them with another form of framing, such as timber or concrete blocks. The bales are then covered in a finishing plaster of earth or lime.

Hemp houses are new to New Zealand, although one was featured on the local version of the *Grand Designs* TV programme, giving a general audience a peek at the technique. 'Hempcrete' combines hemp shiv — chipped woody hemp stalk core — with a lime binder to make a thick slurry that is poured into framed forms in a similar manner to rammed-earth building. The walls can then be plastered or lined. Hemp is renowned for its fast growth rate, although hemp shiv can't be processed for construction here yet, so must be imported.

These forms of building are rare in New Zealand, and need to be permitted as an 'alternative solution' under the Building Code. The Earth-Building Association of New Zealand (www.earthbuilding.org.nz) will be able to point you in the direction of qualified and experienced earth-building professionals.

PREFAB AND MODULAR

Prefabrication refers to any building or building component that is made off-site in a factory and delivered to the site completed or ready-to-assemble. It's also sometimes called off-site manufacturing (OSM). Prefabrication was one of the earliest building forms in New Zealand — Māori used pre-bundled raupō panels to form the walls of their whare, and some European settlers shipped partially prefabricated house frameworks from their homelands. Industry body PrefabNZ points out that prefabrication has a long history in Kiwi residential design, being used in the construction of homes for railway and hydro-scheme workers, in component systems for more than 40,000 Lockwood homes, and in the University of Auckland's modular Elam Hall student accommodation. Around 95 per cent of new timber-framed homes use pre-cut wall frames and prefab roof trusses.

Prefabrication has a number of advantages. It's faster to build off-site (saving time and therefore money — up to 15 per cent of budget, according to PrefabNZ CEO Pamela Bell), there's reduced waste, and reduced exposure of the structural building elements to the weather before a house is closed in with a roof and exterior cladding. Building in a factory is also safer than building on-site, and minimises the chance of chemicals from building materials leaching into the environment.

Prefabrication can be broken into five categories: components (like the widely used pre-cut frames and trusses); single panels that can be delivered complete with insulation and services space; volumes (modules or pods); hybrids (volumes combined with individual panels to make more complex shapes); and complete buildings.

Most prefab in New Zealand is currently componentry, although more building companies and designers are experimenting with the possibilities of panels, pods and modular systems. In August 2017, Fletcher Living assembled a 170 sqm, three-bedroom home at Hobsonville Point, Auckland, in a single day. Compared to the six months-plus that it usually takes to build a standard home in Auckland, the advantages are obvious, especially given the current Labour government's KiwiBuild plan, which aims to address New Zealand's housing shortfall by building 100,000 new homes by 2028.

A key term being used in the prefab market is 'mass customisation'. Put simply, this means modern prefab design-and-build companies work with homeowners to design a customised home using technology-driven components like panels and pods, or a modular building system. This gives the homeowner a greater level of cost and timing certainty than a completely bespoke home, while at the same time enabling a design with a greater level of personalisation than the handful of standard plans offered by group home builders. One challenge that is still being worked through by the industry is that banks can be reluctant to release mortgage finance until a residential building is attached to its site.

Two important construction innovations that are increasingly being used in the prefabrication space are cross-laminated timber and structural insulated panels.

CROSS-LAMINATED TIMBER

Cross-laminated timber (CLT) panels are made of layers of solid-wood boards (usually three, five or seven layers) glued together at 90-degree angles to make them strong enough to be used structurally. They're a fifth of the weight of pre-cast concrete panels, and can be used for walls, flooring, stairs and structural roofing panels. Being wood, they have zero embodied carbon, work as a beautiful interior finish, and have good thermal and acoustic resistance, as well as exceptional performance in an earthquake.

It's a relatively new technology in New Zealand, first emerging in Germany and Australia in the 1990s. In 2012, the first CLT manufacturer in the southern hemisphere, XLam, opened in Nelson, so the panels can be locally sourced, although it's possible to import them from overseas, too.

CLT panels are machine-cut in a factory very precisely and economically to the architect's specifications, complete with holes for windows, doors and services. Panels can be designed to any height or width, as long as they can be transported to site safely. On-site, they are screwed together (and eventually could be unscrewed and reused) to form a very quick structural assembly.

Wellington-based firm Makers of Architecture (who work in partnership with the building company Makers Fabrication) are at the forefront of CLT use in New Zealand. Architectural director Beth Cameron says working with the product allows them to offer quality design environments to more people, while also trying to shift the industry into a more technological manufacturing age.

CLT is a dense product with its own thermal mass, so can be incorporated in passive design, particularly when well insulated. Another benefit is that because it's a natural timber product, the fibres in the product regulate humidity in general living spaces — the building breathes. In a dry, hot climate the timber releases the moisture it holds into the atmosphere, creating a healthy, comfortable interior environment. In a humid climate, the opposite happens: it starts absorbing the moisture.

STRUCTURAL INSULATED PANELS

Also known as 'stress-skin' panels, structural insulated panels or SIPs are typically made up of a rigid polystyrene or polyurethane core sandwiched between two panels — usually oriented strand board, but alternatively plywood, steel, cement board, or in the case of the locally made Magroc, magnesium oxide board. They're more commonly used in the commercial world or multi-unit buildings, but are starting to make more appearances in the single residential market. There are a number of companies manufacturing and distributing SIPs in New Zealand.

Like CLT, SIPs are strong enough to be used structurally instead of framing, and will be precisely custom-made in a factory to a home's specifications, usually including holes for wiring, although plumbing is often placed between the internal lining and the SIP or through foundations to internal timber-framed walls.

SIPs make it very easy to achieve very high R-values in walls and ceilings at a reasonable price, are airtight, and as prefab panels, reduce waste and building time on-site significantly. The rigid foam core means that the insulation is not at risk of slumping or deterioration, and is protected from pests.

As SIPs are commonly made with manufactured timber panels, look for ones with zero-VOC emissions ratings. The foam insulating core typically does not off-gas, but the blowing agent used to produce the foam does produce greenhouse gases and use petroleum products. SIP manufacturers suggest that this can be balanced against the fact that the high insulation value of SIPs reduces energy requirements for heating and cooling (and therefore fossil-fuel use and its greenhouse gas production). It's important to remember though that New Zealand already derives 83 per cent of its electricity from renewable sources, and has a target of 90 per cent by 2025 and 100 per cent by 2035, meaning our homes consume relatively low levels of fossil fuels for electricity compared to most nations.

INSULATION

Ask any healthy home advisor for the first thing to get right when making your home healthier and more sustainable, and they will tell you that it's insulation. New Zealand homes are chronically lacking in insulation, and even those that do have insulation are often not insulated well enough to make them comfortable year-round. On top of that, old insulation may have slumped, shrunk, deteriorated, grown mould or been ripped or otherwise damaged by rats and mice, making it much less effective.

Although we associate insulation with warmth, good insulation actually helps to keep the house cool in summer too. The insulation in your walls, floor and ceiling works in the same way as a Thermos flask does. If you fill the inside with something warm (heat from a heat pump or wood-burner), it keeps it warm. If you fill it with something cold (air-conditioned or naturally ventilated cool air) it keeps it cool.

Like a Thermos flask, insulation has to completely enclose the interior of your home to work its best. If a Thermos has holes or cracks in it, the warm or cool liquid can leak out. And even if those holes or cracks are patched with something, if the patch isn't the same material as the Thermos, it won't work as well.

In your home, the most common 'holes' in your insulation are windows, doors, chimneys and skylights, all of which let heat transfer through them much more easily than an insulated wall does. Ceiling downlights can also be holes, as many of them can't have insulation laid over them as they get too hot and become a fire risk. Some newer LED downlights do stay cool enough that they can have insulation overlaid — it's worth seeking these out.

Insulation value (usually written as an R-value) comes from a combination of all the different materials that make up a wall, ceiling or floor. In a typical timber-framed wall, at a very basic level, there's an inside wall surface, a layer of insulation batting fitted into the timber frame, a layer of external sheathing like a building paper or board sheathing to help keep the building weathertight, and an exterior cladding layer. The internal wall, timber frame, insulation batting, exterior sheathing and cladding all have their own R-values, which are added together to create an overall R-value for the wall.

Trouble is, the timber frame that the insulation batts are fitted into will have a lower R-value than the batts themselves. This means the timber framework is like the patched parts of the Thermos — it'll still keep heat in, but not as well as if the Thermos material was completely continuous. This difference in R-value that allows heat to sneak out is called a 'thermal bridge' — a path of least resistance that heat can take to escape from your home. Thermal bridges often occur at joins between surfaces, like where a wall joins a ceiling or floor, or where a window joins a wall. A good architect will work to create a continuous thermal 'envelope' of insulation that minimises or eliminates thermal bridges.

WHAT'S AN R-VALUE?

The 'R' in R-value stands for thermal resistance, as in how resistant a material is to heat transfer. You will commonly see it expressed to one or two decimal points, such as 0.26, 2.5, 4.0. The higher the R-value, the more resistant the material is to letting heat pass through it — so higher is better.

A timber-framed wall built to Building Code requirements must have an overall minimum R-value of 1.9 in Auckland (Climate Zone 1) and 2.0 in Christchurch (Climate Zone 3). However, healthy-home professionals and sustainable designers advocate much higher R-values for energy efficiency and year-round comfort. In Europe and the United Kingdom, code-standard R-values are up to three times higher than here in New Zealand. Super-insulated Passive Houses typically have wall R-values of 7.0 to 10+.

When choosing insulation material, there are four things to consider — its thermal performance, its ease of installation if you are retrofitting it, its environmental impact, and its durability. Since most heat (30–35 per cent) is lost through your ceiling, that's the first place to make sure you have insulation. Suspended timber floors are next. Windows (21–31 per cent loss) are valuable to address (see the next section) through changing the glazing, frames or window treatments.

If you're building new, you're in the perfect position — building scientist Joseph Lstiburek puts it this way when writing for Building Science Corporation website: 'Whatever you think the right amount of thermal insulation should be, double it and shut up. If you love your kids, don't argue with me.' Aside from the health benefits of a warmer home, good insulation will decrease your power bill, paying for itself over time. A working group from the Ministry of Business, Innovation and Employment has calculated that installing better insulation pays for itself within six years — plus you'll be more comfortable that whole time.

All types of installation need to be correctly installed to ensure there's no thermal bridging or gaps between the insulation and framing, and that it isn't under- or over-packed. It's a job that you can do yourself in some cases, but do take professional advice first, including on which products are best for your house type and local climate.

FIBREGLASS/GLASS WOOL

The classic Pink Batts have been a staple of Kiwi home construction for half a century. Fibreglass batts — now often called 'glass wool' — are made of silica spun into glass fibres, and may contain recycled glass content, which can reduce the embodied energy of the product by more than 50 per cent. Previously, fibreglass batts used binders that contained formaldehyde, but now most are formaldehyde-free. Glass wool is widely available as batts and blankets, has good thermal performance, and is reasonably priced and long-lasting. When glass wool is being installed, it can release fibres that can irritate skin and also may be breathed in, so installation must be done under proper conditions. Glass wool batts can be less effective when damp or very cold. Although there is currently no fibreglass recycling available, batts in good condition can be salvaged and reused.

MINERAL WOOL/ROCK WOOL

Mineral wool or 'rock wool' is spun from molten stone or slag (waste 'rock' left over after metals have been extracted from ore). It is fire-resistant, great at controlling noise and has good thermal performance. It does use binding agents, but low-VOC versions are available. It can be installed as batts or blown in.

WOOL AND WOOL BLENDS

Wool insulation may be made of virgin wool, recycled wool, a blend of the two, or blended with polyester. Its embodied energy is lower than glass wool or rock wool, as it only requires cleaning and degreasing, rather than energy-heavy extraction, but the cleaning process does produce effluent. It is more expensive than glass wool, and has a natural level of fire resistance, though it should be separated and protected from heat sources. Wool insulation may be chemically treated to discourage insects and mould. Like glass wool and mineral wool, pure wool doesn't perform well if wet, though wool–polyester blends perform better. It can be recycled or reused if in good condition, and will eventually biodegrade. Wool is more expensive than glass wool, but wool–polyester blends are about the same.

POLYESTER

New polyester insulation is a petroleum-derived product, so it has a higher embodied energy, although some brands contain close to 50 per cent recycled content from plastic bottles, so look for these. It's widely available as batts and blankets, and has slightly lower R-values than fibreglass at the same thickness, while costing slightly more. Like the previous types of insulation, its effectiveness is lowered when wet. It is safe and comfortable to install. In theory it's fully recyclable, but Level (an arm of BRANZ) says no data exists on local recycling.

POLYSTYRENE

Polystyrene is made from petrochemical products, using blowing agents that create high levels of greenhouse gases, particularly extruded polystyrene (XPS). It has a similar embodied energy to polyester fill, but very good R-values. For retrofitting, it's generally used underfloor, but in new builds it can be installed in walls and ceilings or as an exterior wrap. It's resistant to moisture, but not to fire, so must be protected by a flame barrier. It's recyclable, but it will not biodegrade in landfill.

MACERATED PAPER/CELLULOSE

An option for places where insulation retrofitting is difficult, like tight roof spaces, blown-in macerated paper or cellulose needs to be installed by a professional. It is a biodegradable product (that does deteriorate over time), and often contains a high recycled content. It's unsafe for use around downlights, and must be shielded from heat sources, as it is not fire-resistant, and if it gets wet, it may rot. It's best in dry and windless conditions.

OTHER MATERIALS

Materials that are used for insulation overseas but are less common or not yet available here are cork (a natural, carbon-sequestering sustainable product with good insulating and sound-absorption qualities), aerogel

> ### SAFETY ALERT: BEWARE OF FOIL INSULATION
>
> Foil insulation, most commonly used for underfloor insulation, was banned in 2016 following five reported deaths by electrocution associated with it. If you have it in your home and it is damaged, it cannot be repaired, but it can be removed and replaced with another form of insulation. Always turn off the mains power before inspecting foil insulation, and leave removal and replacement to a professional.

blankets (extremely high R-values, often used to reduce thermal bridging), denim batting (which contains as much as 80 per cent post-consumer denim content) and aerated concrete (a foamed cement insulation with very high R-values, only suitable for installation in new builds).

WINDOWS

New Zealand homes are often designed for indoor–outdoor flow, with big ranch sliders or bifold doors, or large windows to let in light and views. But if that glass is single-glazed, not airtight, or set in a frame that conducts heat easily, it can be difficult and expensive to keep your home warm. Single-glazed windows also attract condensation — many people will be familiar with waking up in the morning to find their windows 'crying' with moisture that leaks onto the frame and sill and can form mould.

A Building Code-standard wall in Auckland has an R-value of 1.9. But a single-glazed window in a standard aluminium frame has an R-value of just 0.15. By comparison, an insulating glass unit (which is basically a double- or triple-glazed window system) with a timber frame, low-emissivity (or low-E) double-glazing and argon gas filling the space between the glass panes has an R-value of 0.51. Compared to the wall, it's still a lot lower, but it's a significant improvement. Add insulating window treatments, and your home will be considerably warmer in winter and cooler in summer.

Replacing or retrofitting windows can be daunting because of the cost of doing a whole house. While you can get some manufacturing efficiencies from saving up and doing the whole lot at once, if you'd rather get started sooner, start with the windows on the coldest side of your house (south or south-west) as these lose the most heat.

Left: The Homestar 10 Superhome uses uPVC window frames, with an inset from the exterior building surface to improve the thermal efficiency. Photo: Juliet Nicholas

Below: When this Titirangi home was renovated, cedar window frames were chosen for their warmth and era-appropriateness. Photo: Simon Devitt

WOOD

The original and the best, in terms of ecological and thermal performance, but also — you guessed it — the most expensive. Many villas, bungalows and even pre-1970s homes have original wooden window framing, but with single-glazing, so need double-glazing or secondary-glazing retrofitted to improve the performance. Wooden windows do need maintenance in the form of staining or painting, but are extremely long-lasting, and look beautiful. Harder woods are more expensive than softer woods, but are more durable.

uPVC

Less common, but increasing in popularity, uPVC (unplasticised polyvinyl chloride) frames have a look designed to imitate wood. They have similar thermal performance to wood at a lower cost, and are very low-maintenance. uPVC windows have been reported to discolour in the sun sometimes, particularly white frames. As with any polyvinyl chloride product, the manufacturing process for uPVC frames has a high embodied energy — 25 times as high as wood — and produces a toxic chemical by-product.

ALUMINIUM AND STEEL

The most common type of window frame in New Zealand, aluminium is lightweight, durable and requires little maintenance. However, it performs very poorly as a thermal conductor unless thermally broken (an interior and exterior framing layer separated by an insulating material), and has a high embodied energy.

Aluminium is extremely energy-intensive to produce — New Zealand's only aluminium smelter, at Tiwai Point near Bluff, is the country's biggest single electricity user, accounting for an incredible 12 per cent of New Zealand's total electricity demand. As a material, aluminium has an embodied energy 70 times higher than wood. Aluminium's main appeal is the price — it's the cheapest type of window frame.

Steel framing is less energy-intensive than aluminium (although still higher than wood, uPVC and fibreglass), but more expensive, and with a similarly poor thermal performance. The strength of steel means it can be used for thinner and more elegant framing profiles.

FIBREGLASS

Newer to the market here, fibreglass framing performs similarly to wood and uPVC, with fewer of the environmental concerns of uPVC, but a similar cost to wood.

AIRTIGHTNESS AND VENTILATION

Insulation, heating and cooling, airtightness and ventilation work as an inseparable system in a home. You can't address one element without thinking about the others.

If your home is not airtight, heat can leak out through gaps in window frames and around door edges, through power sockets and light fittings and even through the walls themselves, wasting the warmth that you've paid for and making your house colder. Cold air and draughts leak in from outside too, making it even harder to keep your home warm. This movement of air through unintentional gaps in a house's structure is called infiltration.

Airtightness is measured in air changes per hour (ACH), and the New Zealand Building Code sets a minimum ventilation standard as one change every three hours throughout the house, with a faster change rate of one change every two hours in the bathroom and one change every hour in the kitchen (where more moist and smelly air accumulates). Most draughty older houses will gallop past this rate, with many having five or more complete air changes an hour (which may be fine in summer, but as noted above, can be chilly and hard to heat in winter).

On the other hand, homes built after 2000 can be in danger of being *too* airtight. Many people don't want to open windows regularly for security reasons, to keep flies and bugs out, or to keep warmth in. In winter, some houses can go weeks or months without having windows opened. But if your house is very airtight, the air can become stale, stuffy and moist inside. Every time you take a shower or bath, cook, use your dryer, water your plants, or even just go around breathing in and out, you're releasing moisture into the air inside your home. If that moist air gets into your wall structure, it can form condensation that literally dampens the effectiveness of your insulation, and leads to mould problems.

Even if your wall structure is protected with an internal vapour membrane (see page 110), the moisture can condense on your linings and ceiling. So you need to let that stale, moist air out of your house regularly and replace it with fresh, outside air — either through a built-in mechanical ventilation system, or by natural ventilation, better known as 'opening your windows and doors'.

Ventilation is a bit like solar gain, in that you don't want too much or too little, but just the right amount. A good maxim to follow is 'build tight, ventilate right.' The best system for you will depend on the age and structure of your house, and your own living habits.

NATURAL VENTILATION

Natural ventilation uses the movement of wind and convection currents (air rising as it warms, and falling as it cools) to cool a building and remove moisture and stale air. It's a widely used technique in tropical countries.

GLAZING

Frames and glass work together as a system, and choosing a better frame can be a waste of money if you don't improve your glass at the same time. Double- and triple-glazed windows use two or three panes of glass, separated by a gap of (ideally) 12 mm, with the gap/s filled with air or insulating argon gas. Triple-glazed windows were rare in New Zealand, but following the arrival of Passive House (which often specifies them), there are now local manufacturers for both timber and aluminium-framed triple-glazed windows. In Auckland and Northland, triple-glazing may be overkill, but in the colder south it has marked value.

There are two main types of glass designed for eco-performance. Low-E (where the E stands for emissivity) glass is coated on one side to reduce heat loss from inside the house. Its 'opposite' is reflective or tinted glass that reduces the amount of heat or light that comes in — which is good for rooms that are at risk of overheating in summer. Other types of glass are designed to break safely ('safety glass') or reduce external noise ('hush glass').

Secondary-glazing is a way of creating a double-glazed window without replacing the original single pane. A pane of glass or acrylic is fitted within the existing window frame (which may reduce sill space) or magnetically attached to a metal frame. This is generally cheaper than retrofitting double-glazing, and performs well, though is aesthetically less pleasing. A budget-friendly short-term solution (lasting about a year) is to use plastic shrink-wrap (3M makes a widely available version), which is taped to the window frame and heat-shrunk-to-fit using a hairdryer.

In cross-ventilation, windows and doors are placed facing the direction of favourable winds, as well as on opposite sides of the room or building, to allow the air to move freely through the space. In stack ventilation, high windows or skylights can be opened or vented to release hot air from the house, while creating an up-draught that draws cool air from open or vented downstairs windows through the house.

Air also cools as it moves across a body of water. When a swimming pool or large pond is placed near the house on the windward side, it can help cool air pass into the house, particularly if the pool is shaded or semi-shaded.

The main issue with natural ventilation is that people just don't do it often enough. In winter they may be too concerned about letting heat out, and in summer worry about letting flies (or burglars!) in. Insect screens and security stays can help alleviate the latter concerns. Trickle vents, or very small vents that can be left open for continuous airflow, can be a good way to keep air fresh, though windows will also need to be opened more widely on a regular basis to keep moisture down.

It's a good idea to fling doors and windows open in as many rooms as possible for 10–15 minutes every day. In winter, you can do this just before you leave the house in the morning. That way you're keeping the heat in until you're about to leave anyway.

MECHANICAL VENTILATION

Mechanical or 'active' ventilation systems can be designed for a specific room (such as bathroom extraction fans, kitchen range hoods or laundry ventilation) or for the whole house. Room-specific ventilation will be covered in the relevant chapters later in the book.

There are two main types of whole-house systems: positive-forced-pressure, and balanced-pressure and heat-exchange systems.

Positive-pressure systems: These systems blow filtered air into the house through one or more ceiling vents, usually drawn from the roof cavity, where air tends to be warmer. While this may dry the house out, it's not fresh air. There's no exhaust aspect to these systems, so the stale air in the house is effectively forced out through open windows or doors, cracks and gaps in windows and door framing, or even into the walls.

Studies by Beacon Pathway, BRANZ and University of Otago researchers have raised questions around the effectiveness and health of these systems. In southern cities, winter roof-space air may be cooler and moister than air already in the house, resulting in no benefit during the winter months. Additionally, if the filters on these systems are not cleaned diligently, they can blow dust, mould and other particles into the house from the roof cavity, which may increase allergies and respiratory disorders. These systems are not suitable for homes with asbestos roof claddings.

Balanced-pressure and heat-exchange systems: A balanced-pressure system with a heat-exchanger draws fresh air from outside the house to fan into the house (usually into living areas), while extracting stale air through other fans, usually placed in the bathrooms and kitchen. The outgoing stale air passes through a heat-exchanger which transfers its paid-for heat into the incoming fresh air. In summer, the transfer can be reversed, cooling the incoming air. The transfer isn't perfect — around 20 per cent of the heat or coolness is lost — but it retains a lot of the heat that you've already paid for, while drying and freshening the air.

Right: Warm air rising into the stairwell void at the Lynds House is naturally vented out through a window at the top. Photo: Makers of Architecture

Below: At the 339 House by Strachan Architects, a lap pool sends cool air moving into the living area through bifold windows. The windows can be partially opened and locked in place as a water-safety measure when children are in the house. Photo: Simon Devitt

These systems work best in houses that are already well insulated and airtight. If your home was built before 2000 and hasn't been retrofitted to improve insulation and airtightness, think carefully before investing in one. Passive Houses often use these systems to achieve the perfect balance of moisture control and fresh air in a very airtight building.

WEATHER BARRIERS AND VAPOUR RETARDERS

Weather barriers and vapour retarders manage the movement of moisture and moist air into a building's structure. Typically, weather barriers go on the outside of a wall or roof frame (behind the cladding) to protect against rain and air that has gotten behind the cladding, and improve weathertightness. Vapour retarders go on the inside (behind the interior lining) to manage moist air diffusing through the lining and into the wall structure and insulation, and improve airtightness.

They can take the form of building papers, wraps or membranes, or rigid sheathing such as plywood boards, and are sealed with building tapes to create a continuous lining. They're usually installed at the time of building, though if you are relining or recladding your walls, they can be retrofitted.

Their main aim is to help protect wall framing from deterioration, protect the insulation from moisture that can reduce its performance, and prevent the growth of mould inside the walls (known as invisible mould, which homeowners may be unaware of) that can affect indoor air quality and increase allergies and respiratory problems. When there is a big temperature difference between the air inside a house and the air outside a house, and moisture is present inside a wall, it can condense on either the interior or exterior linings (whichever one is colder), in the same way that moisture from the air condenses on a cold beer bottle taken from the fridge. If the condensed moisture stays there for a long time, mould can form.

Having moisture inside your walls, while undesirable, is almost impossible to avoid completely, so weather barriers and vapour retarders are designed to let as little moisture as possible in, while also allowing any moisture that does get in to pass back out so that the interior of the walls remain dry.

CHAPTER 7

THE SHELL

The Sawtooth House by Assembly Architects uses an innovative roof design to allow views of the Remarkables mountain range through the high windows, without letting in too much sun that would overheat the house. Photo: Simon Devitt

The exterior skin or shell of your home — comprising the roof, eaves and wall claddings — has three main roles. First, to keep out wind, rain, sun, pollution, pests (and other unwelcome visitors) and noise; second, to help keep warmth in (or out in summer) through adding insulation value; and third, to present an attractive face to the world. In an eco home, the materials used to do this should be durable, energy-efficient, non-toxic and with a low embodied energy. It works in synchronicity with the internal wall structure and the foundations to create the overall building envelope, and a key factor in how well the shell performs is how it is put together with the internal wall and foundation structure to create a continuous weathertight enclosure.

The shell is primarily a response to the climate of a site. Homes built in hot and arid climates traditionally have quite different exteriors to homes built in moist, tropical climates or homes in cold alpine areas. As discussed in Chapter 3: Choosing Your Property, although classed as a temperate climate overall, New Zealand has a wide range of microclimates that need different building responses; a home in Queenstown will need a different exterior approach to a coastal home in Northland.

EXTERIOR CLADDING

The most important role of a building's cladding is to keep the building weathertight. Having said that, almost all cladding systems fail to do this perfectly over time, so the system must also have the ability to allow the building to dry out when moisture does get in. The failure to allow for this was what caused the leaky homes crisis of the late 1990s and early 2000s.

The taller and the more complex a home is, the greater the risk of moisture getting inside the cladding; design features such as minimal eaves, flashings, roof decks and solid balconies can also increase the risk, as can building in a high wind zone; and some claddings are just more susceptible than others to damage and weather ingress. Add-on fixtures such as television aerials and satellite dishes, roof-based solar-power and hot-water systems, poorly installed or maintained guttering and retrofitted skylights can also compromise weathertightness. All these factors need to be considered with your architect or designer when making a cladding and roofing choice.

WOOD

Timber weatherboards, either as a solo cladding or in combination with another form, are a traditional and popular choice that has been around in New Zealand for well over a century. Timber performs well in earthquakes, as it has the ability to move without cracking and crumbling. There are three main weatherboard systems: bevel-back (horizontal with a slight backwards slope, as seen on villas and bungalows), rusticated (laid flat horizontally) and shiplap (laid vertically with an indented profile). Bevel-back is the most air- and weathertight, according to BRANZ.

THE FOUR DS OF WEATHERTIGHTNESS

- Deflection: keeping water away from points of entry through the use of eaves, flashings, facings, sealants and overhangs
- Drainage: providing a way for moisture that has gotten in to drain out
- Drying: incorporating airflow so that retained moisture can evaporate from the wall systems
- Durability: using materials that are resistant to damage and decay

Forestry timber makes a substantial contribution to carbon sequestration, or the removal of atmosphere-warming carbon dioxide from the air; every tonne of wood used in construction has absorbed around 1.7 tonnes of carbon dioxide from the air. That means that choosing timber cladding and ceiling sarking for your home represents around 15 tonnes of carbon sequestration. That's about the same amount as you'd produce by driving your car for 60,000 kilometres.

But there's more to choosing timber than just choosing timber. To be truly eco-friendly, timber should come from a certified sustainably managed plantation, be milled in a way that minimises waste, transported as short a distance as possible to the site, and treated for durability in a non-toxic way.

Some woods are durable enough to be used as an exterior cladding without a form of treatment to increase their strength and water resistance. Locally grown durable cladding timbers are Lawson cypress, macrocarpa, lusitanica, and heart Douglas fir, and Western red cedar and cypresses are the most common imported ones. Softer woods like radiata pine require treating, either with chemicals like copper chromium arsenate (CCA) or copper azoles, or by being strengthened through thermal modification (a heat treatment to decrease moisture absorption) or acetylisation (basically being pickled in strong vinegar). Plywood sheeting is another form of exterior wood cladding. Non-durable timbers need to be treated to improve their durability.

Accoya, the best-known acetylised wood, comes from a company based in the Netherlands. It uses New Zealand-grown radiata pine, which is (unfortunately for its carbon-kilometre profile) shipped all the way to the Netherlands for treatment, then sent back here (and around the world) for use.

Abodo is an Auckland-based timber company that supplies exterior claddings that use thermal modification, naturally durable timbers and the Japanese shou sugi ban 'charred wood' process to make preservative-free timber cladding and decking products.

Hardier timbers like cedar and macrocarpa can be left to develop a naturally weathered patina, while softer ones like radiata pine need to be painted or stained to keep them in good condition.

Reclaimed or salvaged wood is a possibility as a cladding option, giving an attractive pre-weathered look, but it may need extensive preparation to make it suitable for installation, and may present issues with weathertightness.

METAL

After wood, there may be no material more widely used in New Zealand's vernacular architecture tradition than long-run corrugated-metal sheeting. Originally a roofing material (and still the preferred roofing material here), its use has extended to the claddings of all types of buildings, from the

Above: This home at Dalefield, Queenstown, uses Abodo Vulcan Cladding thermally modified timber as an exterior cladding in an area with extreme temperature changes. Photo: Abodo

Right: Shou sugi ban is an ancient Japanese form of weatherproofing that has seen a huge return to popularity in the last few years. Photo: Abodo

classic red barn to modern homes. It can be installed horizontally or vertically (or both). It's available in a wide range of profiles, and is usually a form of galvanised or zinc-coated steel or aluminium alloy. As with metal roofing, the high embodied energy of the material makes it less eco-friendly, although its recyclability, durability and potential recycled content are mitigating factors. Profiled metal sheeting must be applied over an absorbent wall underlay, and drainage, venting and good detailing are important with this kind of cladding.

The Red House by Crosson Architects draws on the vernacular form of the red corrugated-iron shed, giving it a modern twist by rotating panels to create contrast. Photo: Simon Devitt

MONOLITHIC CLADDINGS

Monolithic claddings, which are essentially sheet claddings finished with continuous plaster, are strongly associated with the leaky homes crisis, but in themselves they were not the cause of the problem — it was the fact that they were installed without proper cavity systems for drainage and drying, combined with the use of untreated timber framing, along with the fact that paint was relied on too heavily for waterproofing. Properly installed over adequate cavities and durable framing, monolithic cladding systems can function well.

uPVC

Unplasticised (rigid) polyvinyl chloride cladding (also known as vinyl siding)

EXTERIOR PAINTS AND STAINS

Paints and stains improve the performance of exterior claddings and help extend their longevity. High performance of the coatings is important, so exterior paints and stains can contain ingredients that emit more VOCs than indoor paints. Off-gassing isn't such an issue with exterior coatings, but it's good to look for ones with low toxicity, especially if you're in an area where rain run-off from the building might get into streams, ponds or wildlife habitat.

Lead paint: Many homes painted before the 1980s used lead-based paint. Lead is a serious health hazard, causing brain, developmental and nervous-system disorders, kidney disease and high blood pressure. Microscopic lead particles flake off and accumulate in soil, dust, roof spaces and on surfaces where they can be inhaled or ingested. If you're looking at renovating your exterior paint, and suspect you may have layers of lead-based paint, it's essential to get it tested and take strict safety precautions. Lead dust from sanding can travel on the wind and contaminate surfaces throughout your house — and your neighbours' properties, too. Old windows are also a particular source of lead-paint flaking because of the moving parts — get these checked.

Solar paint: It's not likely to be on the market for a few years yet, but scientists at RMIT University in Melbourne have developed a solar paint that absorbs water vapour from the air and uses sunlight to split the water atoms to create hydrogen fuel. It's proposed to work in addition to solar panels to transform entire buildings into energy-generating machines.

The K Valley House uses salvaged metal as an exterior rain-shield. Underneath the shell, a simple plywood construction was a cost-effective way to create a weathertight, well-insulated envelope. Photos: Patrick Reynolds

is designed to imitate the look of wooden weatherboards. It's extremely low-maintenance and accessibly priced. uPVC cladding comes as branded systems, so must be installed by experienced installers according to the particular manufacturer's instructions. BRANZ notes that they are prone to air-leakiness, fading and increased brittleness from sunlight exposure, which is a particular issue with New Zealand's high-UV sunshine. uPVC's main claim to sustainability is its ability to be recycled at end-of-life.

FIBRE CEMENT

Another cladding material that generally comes as a proprietary system, fibre cement boards are usually made of cement, sand, clay, mineral additives and wood pulp. The cement content (around 40 per cent of a typical board) raises the embodied energy of the product. In the past, fibre cement boards often contained asbestos, so if you are removing old boards from your house, suitable precautions for removal and disposal should be taken. These boards are durable, though do need to be painted or coated as they absorb moisture, and good detailing is critical to keeping them weathertight. They're available as weatherboards or sheets, and can be texture-coated, which requires careful maintenance to stay waterproof. Most locally used fibre cement board is manufactured in New Zealand.

EIFS

Exterior insulation and finish systems (EIFSs) are typically a layer of polystyrene insulation used on the outside of the house, coated with a fibreglass-reinforced base layer and polymer-modified plaster. Polystyrene does have a good insulation value, but it is a petroleum-derived product and off-gasses VOCs.

STUCCO

Stucco is a cement-based plaster finish that has been used since the 1920s — the word 'stucco' is Italian for 'plaster'. It can be applied over a number of substrates, including existing veneers, but cracks need to be repaired quickly or they will worsen over time. Modern stucco is often reinforced with plastic mesh to help prevent cracking. Because stucco plaster contains cement, it has a high embodied energy, but it is long-lasting.

STONE

Stone can be used as a structural masonry material, but it's uncommon; stone installed on panels as a veneer is more popular (and cost-effective). Stone adds a timeless quality to a home, and integrates it naturally and beautifully with the surrounding landscape (particularly if local stone is used), is very low-maintenance and is durable. Stone (and other masonry) veneers typically allow a fair bit of water to get through, so are installed over a wet-cavity system. There are a wide range of stones available for veneer cladding, including local ones like basalt, riverstones, schist, mica, slate, mountain shale and Southern ledgestone.

'Manufactured stone' (or Cultured Stone, which is a brand name) is advertised in New Zealand as an alternative to real stone. This is actually concrete moulded and coloured to resemble a wide range of natural stone types, and can be used only as a veneer, not as a structural material. Manufactured stone is lighter and less expensive than natural stone, but has a higher embodied energy because of the cement content.

BRICKS

Bricks have been made for use in building in New Zealand since the early 1800s; a benefit of our clay-rich soils. Unfortunately, as early European settlers came to learn, unreinforced brick buildings were prone to collapse on our shaky isles. Today, reinforced brick veneer is one of the top two residential cladding choices, alongside wood, for its versatility, wide colour selection, accessible cost and long life. Bricks are fired at high temperatures, giving them a high embodied energy. Using recycled or reclaimed bricks instead of new ones is a possibility, but they have to be selected carefully to make sure that they're still in good shape and have been sourced from buildings that were not contaminated with industrial chemicals.

ROOFS

Roofs are one of the great wasted spaces in cities. Collectively, they form a huge expanse of mostly unused hard surface that adds to the urban heat-island effect (whereby dark paved and tiled areas absorb heat instead of reflecting it, raising urban temperatures), contributes to stormwater run-off and requires regular maintenance to remain weathertight. But there are severals ways to make your roof the crowning glory of your home, rather than an eco 'bald spot'.

LIVING ROOFS

Living roofs are an ancient solution to a very modern problem. Simply, they are roofs designed to grow plants on. A drainage layer is laid on top of a waterproofed and root-resistant roof system, a layer of soil goes on top, and plants are seeded in the soil. Green roofs protect the roof itself, absorb rainwater (reducing stormwater run-off), aid biodiversity by providing habitat for insects, birds and small reptiles, reduce carbon dioxide and add to the beauty of the home, particularly in the case of multi-level or split-level houses where one part of the house overlooks a section of roof, a sleepout or a garage.

They can also improve insulation, keeping the house warmer in the winter and cooler in the summer. A study in Germany suggested that a living roof

Above: Having a structural engineer in the family came in handy when this substantial green roof was incorporated into the design of a grand extension. Xanthe White designed the primarily native garden, which was meant as a gesture to the neighbourhood after the home's green backyard was covered by the house extension. Photo: Xanthe White Design

Right: This South Canterbury home by Daniel Marshall Architects has an extensive turfed roof with native grasses covering it. Photo: Ernie Shackles

Left: A roof deck at this Auckland home by Paterson Architecture Collective provides extra outdoor space on a tight site. Photo: David Straight

Below: The Seascape House by Patterson Architects on Banks Peninsula backs into the hillside and has a grass-turfed roof that blends the building into the landscape while protecting from occasional rock-fall from further up. Photo: Simon Devitt

can give a 3–10 per cent reduction in winter power bills. In addition, they can give a level of sound insulation, which may appeal if your home is under a flight path.

As with many areas of green design, New Zealand lags behind when it comes to the adoption of green roofs. At present, they're few and far between here, while in France, Germany, Canada and Mexico the governments have introduced or are considering planning laws that require certain buildings to have at least part of their roof covered in plants or solar panels.

There are two main types of green roof — extensive and intensive. Extensive are lighter-weight and shallow, usually supporting mosses, sedum and grasses, particularly drought-tolerant ones, and are meant to be enjoyed as a view, not as a garden. Intensive green roofs require substantial structural engineering, but may function as a genuine rooftop garden.

Living roofs work best on flat roofs — anything with a slope above 15 degrees will need retention, but will also have better run-off, so accidental water ingress is less likely. It's possible to retrofit lightweight green roofs, but the load-bearing capacity of the structure will need to be taken into account, and may need reinforcing.

EARTH-SHELTERED HOMES

Made famous first by Australia's searing hot Coober Pedy, where hillside dugout homes are a cool relief from the burning sun, and more recently by the cosy hobbit-holes of Hobbiton, earth-sheltered homes might seem quaint but can be designed in a modern way.

The technique of extending a hillside over a home to give it insulation and shelter is known as earth-berming. It's more suitable for climates with extreme temperatures, so in New Zealand, our rare but beautiful earth-bermed houses are typically seen in the southern regions. It's unlikely to be an option you consider unless you're looking at a site with an extreme slope, but it can be a clever way of making use of an otherwise very marginal property.

ROOF DECKS

Roof decks are a good way of extending living space on a tight urban section by utilising a part of the house that is often ignored or inaccessible. Roof decks require extra structural loading and waterproofing, so generally need to be planned as part of a new build or a substantial extension. They can be gravelled in a Zen garden style to keep them low-maintenance and low-cost, or established as a full roof garden, which will be more expensive, but if the site has no ground-based garden, may give a capital payback on the investment.

SOLAR-POWER SYSTEMS

When you hear the words 'eco home', one of the first images that probably springs to mind is a block of white-framed, blue-black photovoltaic panels on the roof. And while the idea of getting 'off the grid', with its implications of economy, freedom from reliance on a power company, and renewability are appealing, making the switch to solar isn't right for everyone — yet.

Solar systems can be standalone (operating 'off-grid' — most common in rural or isolated areas) or grid-tied (linked to the local power network). Standalone power systems almost always have additional battery storage (as otherwise once the sun goes down, the power's out), and grid-tied systems may or may not have battery backup. In either system, the power generated during the day is first used to supply the house's immediate electricity requirements, which may be more or less than the amount of electricity being generated by the system. If the house's electricity consumption is less than the amount of power being generated, any excess power is either stored in a battery or 'sold back' to the power grid.

Most home systems are 3–6 kW in size, which refers to the maximum amount of power the system can generate at one time. How much of your electricity needs the system covers will depend on the size of your system, how much power you consume overall, and how much use you make of the solar power as it is generated. In general, the more electricity you can use during the day while it is being generated, the more cost-effective the system will be for you.

While it *is* possible to free yourself from the grip of the power companies completely with the right set-up, it's unlikely for most people. Solar power is more likely to be an investment that diminishes your reliance on the grid and reduces power bills for a long-term benefit, rather than one that sets you free completely, particularly if you're fitting a system onto an older home that is less well insulated or airtight (which may have higher power needs for heating).

In addition, not all roofs are suitable for solar panels. They need to be set at an angle of around 20–35 degrees in New Zealand to gather optimum solar energy, so if you have a flat or highly sloped roof, you may need to have them propped (which can add over $1000 to the upfront cost). Large trees and neighbouring buildings that cast a shadow across your roof may also diminish the available sunshine, as can the direction the sides of the roof face. And different parts of the country receive different sunshine hours — houses in the deep South can receive almost 1000 fewer hours a year than in the sunshine capital of Nelson.

If you're considering solar power, you need to do your figures carefully, as the maths of changing in order to save money is complicated, and dependent on factors that could change in the future to make the proposition less or more appealing. The cost of systems, and particularly of batteries that allow you to store power for use later in the day, is still dropping consistently, as competition in the market increases. However, the rate that power companies will buy your excess electricity back at has already dropped considerably,

THE SOLAR ROOF OF THE FUTURE LOOKS LIKE THE PAST

It's not always talked about, but one of the things that can put people off solar power is that most solar panels are — let's be honest — not attractive to look at. Depending on the design of your roof and direction of the sun, the best you can usually do is install them in a less conspicuous location. Fortunately, a solution is coming. US-based renewable power entrepreneur Elon Musk's Tesla Solar Roof is made from solar panels that look like traditional slate or terracotta roof tiles. The tiles went into production at the start of 2018, and early interest in them is strong, especially after Tesla announced that the cost of installing them would be less than the cost of a conventional tiled roof — even disregarding the energy savings. (At the time of writing, Tesla Powerwalls, which form part of the system, were already available in New Zealand, but no date for the roof panels could be confirmed).

The Tesla Solar Roof is made of glass tiles designed to look like a conventional roof. Image: Tesla

Left: Weathered metal panels used for the roof and cladding help this Taiharuru house by Crosson Architects blend into the landscape. Photo: Simon Devitt

Below: The Case Study Villa by William Tozer Architects uses wooden shingles as cladding in reference to their original use as roof gabling. Photo: Carme Arguayo

and may continue to drop — or possibly go up. Some lines companies have started imposing additional charges on customers who generate their own power, effectively creating a 'solar tax'.

Of course, getting a full payback on solar power isn't strictly necessary. We spend money on a lot of things just because we want them and not because we expect them to pay us back — couches, clothes, computer games, holidays and so on. You might want solar power because you love the idea of renewable energy micro-generation, and not so much because you want to claw back every single dollar you spend on it. If you're choosing it from an environmental standpoint, though, keep in mind that getting off one of the other two 'grids' (gas or petrol) may do more for your carbon footprint than switching to solar power. It all depends on your personal priorities. New Zealand already gets a high proportion of its power from renewable resources, and further growth in that proportion is likely to come from government or commercial investment in renewables, not micro-generation.

However, if you live in a location with high sunshine hours, have a suitable roof for solar panels, a well-insulated and energy-efficient home that you intend to stay in long-term (payback time usually exceeds seven years) and can make good use of electricity during the daytime, then solar energy can be a great investment; one that will certainly make you more conscious of your energy use habits, and will likely encourage you to change them in ways that will reduce your overall power usage. The Energywise website (www.energywise.govt.nz) has a solar power calculator that will help you figure out whether solar is a good option for your roof, and how long a payback period might be.

Another option that has recently come to the market is rental solar systems, or 'solar as a service'. A grid-tied power system is installed on your roof by a solar-power company (as long as your roof meets certain criteria) at their cost, and they maintain it for the life of the contract (around 20 years at present). You pay a fixed monthly rate for the service, can use as much of the power generated by the panels as you like, and can sell any excess back to the company. You'll likely also need to buy power from a regular power company to top up the deficit from the solar panels. If you move house, you can take the system with you, or transfer the contract to the new owner. This could be a good option for those who don't want the upfront investment cost of a system, and who can make maximum use of power generated during the day.

SHEET ROOFING

Metal: Long-run metal is the material most commonly chosen for new houses in New Zealand, according to BRANZ, and it's easy to see why — it has many advantages. Fine steel sheeting galvanised with a zinc or zinc-alloy coating to prevent rusting is a durable, affordable and low-maintenance roofing material. Although steel has a high embodied energy, it is almost infinitely recyclable, and some new roofs contain recycled steel content. It's the ideal roof for rainwater harvesting, but has minimal insulation value. Light-coloured metal roofs can reflect heat, decreasing the urban heat-island effect.

Insulated panels/SIPs: Structural insulated panels (SIPs) for roofing are usually two sheets of metal with insulating foam sandwiched between. Some standard SIPs can also be used for roofing. They can give excellent thermal and acoustic insulation, and require little maintenance. However, although the metal in the SIPs may be recyclable, the fact that it has been glued to a foam core makes recycling more difficult.

TILES

Wood: Timber tiles are known as shingles (smooth-surfaced, with a tapered shape) and shakes (squarer, with a textured or raw look to the surface). They need a steeper incline than many other products (at least 18 degrees, and ideally 25–30 degrees, which is steeper than the standard 'quarter pitch' of 22.5 degrees), and in damper climates may last less than 10 years, which is why we don't see them commonly in New Zealand. As a sustainable material, they're one of the best, particularly if you source high-quality local tiles, although many shingles and shakes are made from imported cedar. They need good airflow in order to be able to 'breathe' and dry, or their lifespan can be reduced. Like all wooden cladding products, they need maintenance to last, and should be resealed periodically, as well as cleaned. Some warping and change of colour is common, though many people like the natural aging effect.

Slate: Slate tiles are made from thinly split stone. Slate can be sourced from New Zealand, but is typically imported from overseas, creating a large carbon-kilometre footprint. The most sought-after slate comes from Wales. It looks beautiful, but is very heavy and fragile so is difficult to install, requiring a specialist roofer. Its advantages are that it's a natural material (although quarrying has environmental impacts), is minimally processed, low-maintenance and resistant to rot and corrosion, and is extremely long-lasting (centuries, in some cases). Reclaimed slate is increasingly an option.

Clay: Clay or ceramic tiles are normally made from terracotta, hard-baked to a very high temperature. They're durable and low-maintenance, and add a layer of noise protection to the house, so are suitable for high-density neighbourhoods. They shouldn't be walked on as they're prone to breakage, but in the event that one or more are broken (perhaps by falling tree limbs), they are easier to replace than sheet roofing. They need a pitch of at least 10 degrees. They don't rust or warp, and can prevent heat gain to a home.

They do have a high embodied energy owing to the energy used in firing them, and also their weight, which adds to the cost of transport.

Concrete: Similar to clay tiles, concrete tiles are individually pressed tiles made from concrete, which has a high embodied energy and releases carbon dioxide in the manufacturing process. Although solid and heavy, they usually don't have internal reinforcing, so can be broken if dropped or stepped on. They require more framing and load-bearing capacity than metal products, which adds cost to the build. Concrete tiles have a lifespan of at least 50 years, so can potentially be sourced second-hand, recycled after use, or can be crushed to use as fill in new concrete.

Metal: Pressed-steel tiles are more lightweight than concrete or clay tiles, requiring less structural support, and costing less to transport. In an average-sized home this could save you a few thousand dollars, which you could invest in other sustainable elements. They are usually painted, or have a textured coating to make them resemble wooden or concrete tiles. Some pressed-steel tiles feature geometric patterns, and come in sheets that are pressed to look like several tiles, meaning installation can be done more quickly. They're particularly suitable for earthquake-prone areas. Like other steel or aluminium roofing, they can be recycled at the end-of-life, and new metal tiles may contain 30 per cent or more recycled content.

Copper: Copper tiling and panelling is very beautiful, with a natural patina that develops over time. Significant copper roofs can be seen atop the Holy Trinity Cathedral in Auckland, Auckland Museum and Dilworth School Hall, and there is increasing domestic use, particularly for guttering and seams. Although metals have a high embodied energy cost in terms of extraction, transport and production, copper is one of the less 'expensive' metals to extract and process. It also has a long lifespan (100+ years) and is fully recyclable, meaning it comes out well in a full life-cycle assessment. However, the high cost means it's unlikely to be a viable option for most builds.

Asphalt: These shingles are not a structural tile in the same way as other tiles, but act like a veneer on top of a waterproof plywood substrate. Felt material is impregnated with asphalt (a sticky black petroleum product) then coated in more asphalt. A topcoat of crushed rock is used as the finish. The shingles are considerably lighter than concrete and clay tiles, and in some circumstances can be applied over an existing roof rather than replacing the entire roof. As asphalt is a petroleum product, these shingles are inherently environmentally unfriendly, although they can be recycled for use in roading repair after use. They can also off-gas VOCs, and there is concern that some may contaminate rainwater harvested for drinking. Check this, as well as options for post-life recycling with the product manufacturer.

Plastic composite: An alternative to slate is a type of tile marketed as 'synthetic slate', which is in reality plastic composite tiles made from 80 per cent recycled compound that includes rubber and polymers from car bumpers, radiator hoses and door mouldings. It's half the weight of slate (so may be a better alternative in earthquake-prone areas), has a high impact rating (the material is flexible) and is UV resistant. Manufacturers often claim that these tiles are 100 per cent recyclable, although this may be more theoretical than practical, owing to the difficulty in separating the components of the composite. Again, as it is made from petroleum, there is an inherent level of environmental unfriendliness in the product, even though it is recycled.

RAINWATER HARVESTING

New Zealand is fortunate to have abundant rainfall in most parts of the country, although in coming decades, climate change may decrease this, or change the way the rain falls — from smaller, more consistent amounts to less frequent deluges.

In some areas, tank water is a necessity for survival, but in most parts of the country harvesting rainwater is a significant yet relatively easy way to make use of an abundant free resource while lightening the load on the stormwater system. Except in areas where rainwater is the only source of water (and is treated accordingly), harvested rainwater is usually considered non-potable and suitable only for use in washing machines and toilets and for gardening.

Fortunately, apart from bathing, those three uses account for a large portion of our water use. According to Beacon Pathway, the average daily water consumption for use indoors and out is 241 litres per person, but only 5 litres of that needs to be suitable for drinking and cooking.

The size of tank you choose depends on what you want to use it for and how much rainfall you get. In Auckland, where there is regular rain, a 240-litre rain barrel might get you through a summer's worth of garden watering. In Canterbury, the same size would be an emergency backup only.

A slimline 5000-litre tank should be large enough to supply most of your water for gardening, and in rainier areas could be enough for most of your laundry and toilet flushing, too. That size of tank will fit on a space a little less than 3 m long x 1.2 m wide, and will be a little under 2 m high (about the size of an SUV). If you're building a new home, it's possible to bury water tanks underground, but this will likely require the advice of a structural engineer, and an investment in concrete tanks.

Water weighs in at a hefty 1 kg per litre, so a 1000-litre tank literally weighs a tonne. If you want to use a gravity-fed system to distribute the water across your garden, the tank will probably need to be elevated, meaning it will need to be placed on a structurally secure base. If you have a sloped section, you may be able to get away without this. Adding a pump to your system increases the cost and brings with it maintenance requirements. You'll also need to work out where any overflow is going to go — whether it will drain off on your property or be piped into stormwater drains.

Above-ground tanks are usually installed underneath the eaves of your house, below a downpipe, which can be easily adapted by a plumber (or a competent DIYer) to vent into the tank. A filtering mesh is essential to keep out bits of debris, leaf litter and even bird and animal droppings that can end up on your roof and find their way into your tank.

If you're installing a tank just for garden watering, you probably won't need a building consent, but if you're connecting it to your house plumbing then you will. Tanks above 6000 litres may require resource consent. Always check the council regulations around rainwater tanks for your area before installing one.

BREAD

CHAPTER 8

THE KITCHEN

This Wairarapa eco home uses low-impact materials throughout, including Quartzite benchtops. Photo: Elizabeth Goodall

As the heart of a home (and the best place to be at a party), the kitchen is a hub of daily activity. And all that activity uses energy, water and — more so than in any other part of the house — creates waste. It's a wet area, so it needs good waterproofing and ventilation, and it needs long-lasting, durable materials for the working surfaces that can have high embodied energy. That makes it a great place to start when you're trying to decrease your environmental footprint. Whether you're looking at a serious renovation where big, long-term wins are possible, or small ways to make changes that add up over the years, the kitchen offers a real buffet (sorry!) of possibilities.

CABINETRY, BENCH-TOPS AND FITTINGS

Kitchens are a complicated jigsaw to put together, so lots of thought and planning around how you use your space is essential. The design axiom 'form follows function' is particularly applicable — how you like to use your kitchen and the functional priorities of safety and efficiency will dictate how the basic room takes shape.

When it comes to kitchens, bigger is not necessarily better — in fact, it may well be worse. If you're thinking of expanding your kitchen, ask yourself whether you really need that space for functionality or whether you've been swayed by the aesthetics of the sweeping kitchens you see in magazines. Many of those beautiful kitchens are lightly used — a 2014 survey by market -research company GfK found that in countries like Australia, the United States and the United Kingdom, on average people spent just six hours a week preparing meals at home.

Open-plan kitchens have become a standard part of modern living design, often with only a large, standalone island to indicate the separation between kitchen and living/dining spaces. But it hasn't always been that way. Kitchens used to be regarded as 'machines for cooking' — functional spaces where food was prepared as efficiently as possible, and it wasn't until the 1960s that open plans started becoming more popular and kitchens started being seen as social spaces.

However, an open space isn't always great for kitchen functionality. Cooking releases moisture, odours and combustion particles and gases that need to be removed, and the larger the kitchen, the bigger the airspace that needs to be cleared. Powerful modern range hoods might do the job, but they also suck out large amounts of air that you've paid to warm in winter. Plus, kitchen materials are expensive and when you're buying by the metre, keeping your kitchen efficiently sized lets you invest in higher-quality materials.

TIPS FOR AN EFFICIENT KITCHEN LAYOUT

- Every kitchen has a 'working triangle' of sink, stove and fridge. The sink is the point that needs greatest access, followed by the stove and the fridge. If you map the triangle on your kitchen plan, the total length of the three sides should be no less than 10 feet (or it will be hard for more than one person to get around in) and ideally no more than 25 feet (or you will be carrying things too far).

- Think about which doors and drawers open most frequently (dishwasher, oven, fridge, pantry) and whether they impede traffic flow or other regular access points when opened. Roll-up doors on some cupboards may be more practical than doors that swing open.

- Leave at least 40 cm of clear bench space next to your cooktop and fridge as 'delivery space' for food items being removed from them.

- The more you like to cook, the more bench space you're likely to need.

- Store items for easy access — pots, pans and frequently used oils and seasonings near the cooktop, knives on a magnetic strip near the food prep area.

- Insufficient storage is one of the biggest problems in a kitchen. Look for every opportunity to fit more storage into wasted spaces. Building cabinets right up to the ceiling gives you space for items you don't need to access often (and means you don't need to dust on top of them). Pull-out pantries are a great way to fit a lot into a small space.

- Let in the light. Good light is essential in a kitchen. Look for ways to maximise light over working spaces — big windows, tracks of directional lights directly over benches (you don't want your shadow to fall on the bench) and strips of LED lights underneath wall-hung cabinetry.

- Choose an island, not a continent. Kitchen islands have crept up in size over the years, leading to wasted space in the centre. Make your island do double-duty by integrating a breakfast bar.

CABINETRY

Modular (or kit-set) kitchen cabinetry first came to the market in the 1930s. Since then, it's been the standard way of fitting out kitchens. Modular kitchen systems are the least expensive form of cabinetry, and often use chipboard, fibreboard or plywood covered with a melamine surface. Composite wood products are made with resins and adhesives that can off-gas VOCs. If you have a standard modular kitchen where the layout works for you but the doors and handles are looking tired, you can save on doing a full renovation by simply replacing the doors and handles.

More expensive semi-custom or custom kitchens are more likely to be constructed in solid wood, aluminum or steel. Ask for the VOC-emission information of the cabinetry you are considering, and look for low- or zero-VOC products.

Using recycled cabinetry is another environmentally friendly option. Salvage cabinetry can be found at demolition yards and on Trade Me, although you will be much more limited in what will be appropriate for your kitchen. But with the right reconditioning (solid wood can be sanded back and repainted or refinished using low- or zero-VOC finishes and a fresh set of handles), this could be the most cost-effective option.

BENCH-TOPS

Bench-tops are one of the most expensive and avidly discussed aspects of a kitchen renovation. The most 'aspirational' surface tends to be granite or marble, which are renowned for their durability as well as their beauty, but despite stone being a long-lasting natural product, it's not necessarily the greenest choice.

Laminate: Ubiquitous in the 1960s, Formica and other high-pressure laminates (HPLs) were first made to replace tile and linoleum surfaces. Today, they're less common but still available in a wide range of finishes and colours. Laminate is made from layers of resin-impregnated paper glued to a composite board backing. If the composite board and glues used are low-VOC, this is a low-cost, moisture-resistant and stain-resistant option. If you're looking for durability, however, it's not the best choice — it can be chipped, scratched and burned, which may result in an earlier need for replacement than other surfaces.

Bauxite solid surface: As with high-pressure laminates and Formica, the brand name Corian is often used as a catch-all term for a solid surface made from alumina trihydrate (derived from bauxite ore) mixed with acrylic or polyester resins.

Bauxite mining is a hot environmental issue in Australia, the world's largest producer. Tropical rainforests are clear-felled to strip the land for mining, vast amounts of water are used to process it, and the process leaves

The ceiling-hung wooden benchtop at the Gittos House by Spacecraft Architects uses macrocarpa sourced from a felled wind-break. Photo: David Straight

a caustic iron oxide waste product colloquially known as 'red sludge' that is problematic to dispose of. An accidental spillage of a million cubic metres of red sludge in Hungary in 2010 caused chemical burns to more than 80 people and wiped out all life in the Marcal River downstream of the spillage. This type of solid surface is low-VOC and non-toxic, although they can be scratched, stained and are not very heat-resistant. As a solid material, however, it can be sanded back to remove light scratches (even by the homeowner) or professionally refinished.

Quartz composite: Also known as engineered stone, this solid surface is made of 90+ per cent pulverised quartz dust or chips bound in a resin to form a continuous solid surface that is inert and highly durable. Quartz is a by-product of several mining processes, so this is sometimes advertised as a 'green' choice. It's extremely durable, and unlike stone does not need sealing, which can give you some of the beautiful qualities of stone or marble without the maintenance requirements (and it's less expense). It offers design flexibility when it comes to edgings, unlike laminate. Some brands of quartz composite bench-top carry Greenguard certification.

Stone: Although granite is the best-known bench-top stone, there is actually a huge range of options available, including travertine, marble and sandstone, in a variety of colours, patterns and degrees of hardness. Stone is sourced from all around the world, accumulating carbon kilometres as it is transported. When stone comes from less-developed nations, like Brazil or India, the standards of health and safety for the quarry and stone workers may be compromised — cutting stone releases respirable silica particles that can cause silicosis and lung cancer. Although granite has become cheaper than it used to be because of efficiencies in manufacturing and transport, if the stone you're looking at seems to be surprisingly well priced, it's worth investigating where it has come from out about and finding the worker standards there. If possible, look for locally quarried stone or stone with a transparent sourcing process.

Stone is undeniably beautiful as a working surface, and gives a tactile connection to nature. It's naturally water-resistant, inert and emits no VOCs, though needs to be sealed to prevent bacteria and microbes from accumulating in tiny fissures in the surface. Some stones, especially lighter ones, can be stained or damaged by acidic liquids like lemon juice or vinegar, so be sure to test a sample piece thoroughly before committing to it.

Metal: Stainless steel has been used as a kitchen surface for a long time, and it has an industrial or commercial-kitchen quality to it. Easy to clean, hygienic and resistant to corrosion, like all metals it has high embodied energy but is a long-lasting surface that is recyclable and often has recycled content. Salvage stainless steel is an option if you are replacing a standard-sized kitchen.

Above: Quartzite bench-tops, like the one at this Wairarapa home by Donnell Day Architects, use pulverised quartz dust in a durable resin. Photo: Elizabeth Goodal

Right: Stainless-stee bench-tops make a hard-wearing contrast to the extensive interior wood and colour pops of this Titirangi home by Dorrington Atcheson Architects. Photo: Emma-Jane Hetherington

BREAD

Stainless steel does scratch, but brushed versions are now available. Brushed aluminium and copper are other options.

Concrete: Another industrial-looking option, concrete gives flexibility in the bench-top form and edging, can have colours or different aggregates added to suit your décor, and a range of finishes from raw to highly polished. It has a high embodied energy, and the production of cement releases significant greenhouse gases, though concrete bench-tops can contain less cement than structural concrete. Concrete bench-tops can crack and may need to be reinforced to prevent this, and should be sealed for protection.

Solid-surface terrazzo is similar to concrete in that it contains a marble or granite-chip aggregate in a cement or resin binder. Most epoxy resins contain bisphenol A (BPA), an organic synthetic compound that has been banned for use in baby bottles in several countries due to concerns that it may be an endocrine (hormonal) disruptor. If you avoid BPA in your food-storage containers, you may not want it in your bench-top (or flooring) surface either. Look for terrazzo that uses polyester resins instead.

Wood: Solid wood or butcher's block (strips of wood glued together to make a solid composite) bench-tops are a naturally beautiful surface that can be used to prepare food on directly when unsealed. Like your chopping boards, they will scratch and burn if not treated carefully, and must be cleaned regularly to prevent the growth of mould and bacteria, although wood does have a level of natural anti-microbial resistance. Wood bench-tops can also be sanded back and refinished easily. Wood absorbs moisture, and can warp and swell, so is best when used away from the sink area. Salvaged timber is an option — look for durable hardwoods. For new products, look for those that are FSC- or PEFC-certified.

Other less common bench-top surfaces are glass composite, paper composite, bamboo, cork and porcelain.

TAPS AND MIXERS

Kitchen sinks can chew through water. Many new kitchen mixers are fitted with low-flow water restrictors or aerators that can get the flow down to as little as 2 litres a minute (though 4 litres is usually more practical). Otherwise, post-market fittings can be bought from hardware stores.

Two of the biggest culprits when it comes to water wastage in the kitchen are rinsing dishes before putting them in the dishwasher, and using waste-disposal units. Around a third of households have an in-sink food-waste disposer that they use to get rid of a portion of their food waste — a practice that has been criticised as using large amounts of water, and making wastewater more difficult and resource-intensive to treat.

THERE'S SOMETHING IN THE WATER

Heard of 'plumbosolvency'? Didn't think so. It's a charming-sounding word that means something less than charming — the ability of a liquid (like the water in your kitchen and bathroom plumbing fittings) to dissolve other substances that it comes into contact with, such as metals.

If water is slightly acidic (which, according to Watercare, most tap water in New Zealand is), it is capable of dissolving metals like lead, nickel, cadmium and copper from the plumbing fittings in your home. This is a concern, because toxic heavy metals, particularly lead, dissolved in water and consumed, can build up in your body and lead to long-term health effects, like nerve, kidney and brain damage. The Ministry of Health recommends that all households flush a glass of water from their taps every morning. So don't put your kettle straight under the tap to fill it for your morning tea or coffee — let the water run for three or four seconds first. To avoid wasting it, put a watering can underneath to collect it for your houseplants.

APPLIANCES

Between refrigeration, cooking and washing dishes, kitchens account for around 17 per cent of a household's energy use. The biggest chunk of this is in the fridge-freezer, which uses about 10 per cent of the total energy — mainly because it's on all the time. Cooking (stoves, ovens and small appliances) uses about 6 per cent, and dishwashers around 1 per cent, mostly to heat the water up at the start.

FRIDGES

Fridges are one of the longest-term appliance purchases that we make, so it's worth comparing models carefully. Like other white goods, fridges all carry an energy rating — more stars equals better.

Old fridges need to be disposed of carefully. All fridges contain refrigerants (hydrofluorocarbons used for cooling) that need to be removed for reuse so that they don't escape and contribute to global warming. This is a much more widespread issue than many people realise — in his book *Drawdown*, author Paul Hawken identifies worldwide phase-out and disposal of hydrofluorocarbons (which are also present in heat pumps and air conditioners) as the number-one solution to global warming. Contact your local council for recommendations for fridge disposal in your area.

OVENS AND COOKTOPS

Many food-lovers aspire to a gas cooktop because they're seen as giving a more professional level of heat control, and depending on where you live, they may be a more economical way to cook than the traditional electric cooktop. But if the environment is a priority for you, induction cooking is the clear winner.

Induction cooktops work by running an alternating current through a metal coil in the cooktop to generate a magnetic current. When there's no pot on the cooktop, it's cool to touch even when turned on. Once you put a pot made of a magnetic metal like stainless steel or cast iron on it, the magnetic field induces an electric current within the pot itself, making it heat up very quickly. Induction is faster and more efficient to heat than gas — which in turn is faster and more efficient than standard electric — as no heat energy is wasted in the gap between heat source and pan.

The other big advantage of induction and electric over gas is that they don't release as many combustion products as gas. Burning gas releases carbon monoxide, nitrogen dioxide and formaldehyde, which can deteriorate inside air quality significantly. Induction cooktops are certainly more expensive, starting from around $1500 on special (while ceramic cooktops start from $400 and gas from around $800) but with greater energy efficiency than ceramic cooktops and none of the air-quality issues of gas, this one's worth the investment.

The Ministry of Business, Innovation and Enterprise's Smarter Homes site (www.smarterhomes.org.nz) advises that convection ovens are more efficient than standard ovens, and ovens that have a self-cleaning function (where the oven can be heated up to very high temperatures that carbonise all food particles and residues, which can then be easily wiped away) are more efficient on a day-to-day basis because of their higher levels of insulation.

HOW TO FIND THE MOST EFFICIENT APPLIANCE

The Energywise website (www.energywise.govt.nz) has a brilliant running-costs calculator that will help you compare the annual energy costs of different appliances based on their star ratings. The differences can be quite significant — for example, a 400-litre fridge-freezer with a 2-star energy rating might cost you $138 a year in running costs, but if it had a 4-star rating, it would cost only $81 a year to run. That's a $57 annual difference. Investing a little more in a higher-rated appliance upfront may have significant cost savings after the payback period.

TIPS TO SAVE ENERGY IN YOUR FRIDGE AND FREEZER

- Don't make it too cold. According to Beacon Pathway, the ideal temperature is 2–4 degrees Celsius for the fridge, and -15 to -18 degrees for the freezer.
- If your fridge or freezer has coils at the back, vacuum them annually. Dust build-up makes them function less efficiently.
- Forgo the ice-maker. They can increase the energy consumption of a fridge by as much as 20 per cent.
- Keep it cool. Don't place it where it's exposed to heat from a window or skylight, or next to an oven.
- The ideally stacked freezer is completely full, and for a fridge, three-quarters full for maximum efficiency.

RANGE HOODS

Range hoods are available as overhead and pop-up 'down-draught' systems that fit into your bench-top. The most important aspects to consider with a range hood are that it's sized appropriately for your needs, that it's placed close enough to your cooking surface to extract the air efficiently, and that it's properly vented to the outside of the house. You'll also want to look for one that's easy to clean and quiet to use. More powerful range hoods are more effective at removing steam and odours, but they also suck heated air from your home in winter, bumping up your power bill. Turn them off as soon as you've finished cooking.

When installed as recirculating ventilation, range hoods pass the extracted air through a carbon or charcoal filter and send it back into the kitchen. This removes smells, grease and certain combustion particles if you're grilling, but it won't do anything for moisture, carbon monoxide or nitrogen dioxide produced by burning gas if that's your cooking method.

Installing a range hood with ducting to the outside (not the roof space — doing this is now illegal as it's a fire risk, so if your existing range hood vents to the roof space, you should get it changed) is more expensive than installing it as recirculating, but recirculating offers so little genuine environmental benefit that unless it's your only option due to living in an apartment, it's better to wait until you have enough money for outside ducting and do it then.

RECYCLING: WHAT GOES WHERE?

Even diligent recyclers get confused about what goes in which bin (and what can't go in bins at all), and it doesn't help that rules about what's acceptable in council bins differ from region to region. This is not an exhaustive list, so check with your local council if you're unsure about something.

Council recycling (for weekly curbside collection)

YES: Plastic drink and food bottles and containers, shampoo and cleaning product bottles, glass bottles and jars, Tetra Pak cartons (for juice, dairy products and cooking stock), steel and aluminium tins, cans and empty aerosols, all paper and cardboard, newspaper, magazines.

NO: Plastic supermarket bags (if you use these to carry recycling out to the bin, empty the contents into the bin and put the bags in your soft-plastic recycling), tiny plastic scraps like bread tags, broken glass, Pyrex, lightbulbs, window glass, mirrors, metal appliances, plastic toys and furniture, clothing, polystyrene or nappies.

Soft-plastics recycling (Love NZ bins found at various supermarkets and The Warehouse)

YES: Any easily-crumpled plastic food storage bags, including ones for pasta, rice, bread, wraps, frozen food, lollies, cereal, sugars etc., supermarket carrier bags and produce bags, plastic-netting fruit bags, toilet-tissue and paper-towel wrap, silver-lined packaging for crackers and chips, biscuit packets, muesli-bar wrappers, ice-cream and chocolate-bar wrappers, pet-food bags, dry-cleaning bags, bubble wrap, protective plastic sheeting, courier bags, magazine- and newspaper-delivery bags.

NO: Plastic bags designed to biodegrade or compost.

Compost (in your own backyard)

YES: All vegetarian scraps except eggs, but including tofu, tea and coffee grounds, shredded paper (just not too much), rinsed and crushed eggshells, grass clippings, soft 'brown' waste like leaves or cabbage tree fronds, used paper towels, bamboo skewers and toothpicks, pure cotton balls, dryer lint, used matches, untreated sawdust, feathers, latex balloons, chopped-up wine corks.

MAYBE (depends on your system): Dairy products, meat, fish and egg scraps, compostable nappies (wet only, not dirty) with the plastic waistband removed, compostable plastic bags, cooking oils.

NO: Large bones, hard garden waste like tree branches and palm fronds, flax.

Garden waste (private collections, usually on a monthly hire basis)

YES: Weeds, lawn clippings, leaves, palm fronds, branch prunings, small plants.

NO: Noxious weeds like wild ginger, bamboo, flax, branches thicker than wrist-size or longer than 1 m, dirt, clay.

Inorganic (for annual curbside collection)

YES: Large and small appliances, electronics, tools, fixtures and fittings, furniture, sporting equipment.

NO: Building and construction waste — particularly anything containing asbestos, gas cylinders, paints, petrol and chemicals, tyres and car parts, hazardous materials.

Landfill (for weekly curbside collection)

YES: Any small, regular household waste that can't go into the above bins or collections.

NO: Paints and solvents, hazardous chemicals, fluorescent light-bulbs, batteries, medical waste, building materials. These items need specialist disposal — contact your local council for advice.

Other public collection systems (keep an eye out)

Clothing, books, toys, and homewares (blue community bins, Salvation Army, local church missions). Oral care (schools and some dental clinics; see www.terracycle.co.nz). Yoghurt and tuna pouches (by courier box; see www.terracycle.co.nz).

DISHWASHERS

Modern dishwashers aren't just convenient — ones with a 5-star WELS rating use as little as 7 litres of water per wash, making them more environmentally friendly than washing dishes by hand. Many have an eco-mode that can further decrease the energy used to wash a load by 40 per cent or more. Wait until your dishwasher is full before putting a load through — many appliances use a similar amount of water for a full load as they do for a half load.

WASTE AND RECYCLING

It's not hard to understand why the kitchen is the biggest ongoing source of waste in the home — just open your fridge and your pantry door and you have the answer right in front of you. As the hours we spend preparing fresh food in our kitchens goes down, the amount of pre-prepared and packaged food goes up, and with it the amount of packaging we throw out.

According to Paul Evans, CEO of the Waste Management Institute of New Zealand (WasteMINZ), Kiwis generate around 15.5 million tonnes of rubbish every year — that's 3200 kg of rubbish for every person, every year. And we recycle only about 28 per cent of it, leaving 11 million tonnes to go to landfill every year.

Not all of that waste is household waste — in fact, a large chunk can be accounted for by industrial, commercial and construction waste (which, of course, we're still responsible for indirectly). Over hundreds or potentially thousands of years, this waste gradually decomposes, generating high-level greenhouse gases like methane, which can also seep into groundwater and soil.

We're slowly getting better at waste management and recycling, and our country has ambitious waste goals — Auckland and other regions around the country have a goal of zero waste to landfill by the year 2040. Many waste-minimisation experts agree that around 65–75 per cent of the stuff we throw in landfill right now does not need to be there, and with better stewardship can be eliminated.

We all have a natural tendency to do something the easiest way. If you have two options, and one is a lot less hassle than the other, you're far more likely to choose the easier one. Recognising this tendency and designing our environment to support good eco-habits instead of bad ones helps us reduce waste and recycle more effectively.

DESIGNING A WASTE-FREE KITCHEN

Audit your eating: Most of us are habitual eaters. Pay attention to the foods you regularly eat over the course of a couple of weeks or have stored in your pantry, fridge or freezer, and think about the packaging they come in and how you dispose of it. Tiny changes to your shopping habits can easily eliminate rubbish that you throw out every week. For example, if you buy chicken breasts every week, taking a reusable container to the supermarket's butchery section rather than grabbing a pre-wrapped tray from the chiller will eliminate 52 pieces of plastic wrap, 52 plastic trays and 52 of those disgusting chicken-juice soaked pads from your rubbish every year. Make a marinade when you get home and add it to the chicken container, and you're ahead of the game.

Stick to fresh ingredients: There's a direct correlation between the amount of processed food you eat and the amount of rubbish you generate. Try to shop the periphery of the supermarket — that's where all the fresh food in the least packaging is kept — and avoid going down the aisles. Or shop at farmers' markets — fresh foods without the temptations of their convenience-packaged cousins.

Learn to make basics: There is a huge amount of pleasure and accomplishment in making your own staple foods like bread, butter, yoghurt, stock or pasta sauce — and it's usually a lot easier than you think. Wraps and tortillas can be made in just a few minutes in a tortilla iron, a block of fresh butter can be whipped up from cream and a pinch of salt in less than two minutes, and a month's worth of pasta sauce can be jarred in an afternoon using just four ingredients — no plastic bags, Tetra Paks or tins required.

Glass, ceramic or plastic storage jars (especially recycled or second-hand ones) are a great way to store bulk foods.

Plan your meals and shop to a list: In the book *Drawdown*, Paul Hawken identifies reducing food waste as the third most significant thing we can do to combat global warming. While the majority of that waste needs to be eliminated at the front end — growers and supermarkets throw out produce just because it's slightly misshapen or bruised — in high-income economies, Hawken writes, up to 35 per cent of food is thrown out by consumers.

According to a report prepared for the Ministry for the Environment, every year the average New Zealander sends 61.2 kg of food waste to landfill. Meal planning helps decrease that waste (and the chore of having to regularly expunge those soggy cucumbers and the expired meat that you didn't get around to coming up with a recipe for from your fridge).

Look for opportunities to use ingredients in multiple meals — like half a cauliflower in a macaroni cheese one night, and the other half in a soup a few days later. Dana Gunders, author of *The Waste-Free Kitchen Handbook*, recommends planning at least one 'lazy meal' a week — something that you can just pull out of a freezer, or don't have to think about too hard.

Eliminate disposable bags: You probably already bring your own reusable carrier bags to the supermarket, and the major supermarkets are slowly starting to eliminate plastic carrier bags from checkouts as well. Lightweight reusable produce bags (usually made of string or recycled plastic mesh) will help bring your plastic bag count to zero.

Invest in reusable storage containers: Reuse jars or buy storage containers for baking ingredients, spices and dried foods like pastas and grains that can be refilled at bulk food outlets. Some bulk food stores even have online shops that will send your order straight to your door in paper bags.

Segment your bins: Segmented under-bench bins make it easy to separate your different kinds of waste. Ideally, you'll have four: one for 'regular' recycling, one for soft plastics, one for compost scraps and one for landfill. Size-wise, you may find it works best to make the domestic recycling bin the largest one and the landfill bin the smallest, with soft plastics and compost scraps in the middle. Companies such as Hideaway Bins can retrofit under-bench bin systems to your particular kitchen space.

Grow your own herbs: Besides being expensive, store-bought herbs almost inevitably come wrapped or boxed in plastic. Save money and waste by growing potted herbs on your counter, windowsill, balcony or deck (close to the kitchen is best). The trick with herbs is to plant each type in large, individual pots. This stops aggressive herbs like mint or oregano out-competing less aggressive ones like basil or parsley. Growing them in large pots means you'll always have more than you need, and won't pick your plants clean every time you cook.

CAN I KEEP MY COFFEE-POD HABIT?

A waste challenge that has been increasing over the last five years is the rise of the single-use plastic or aluminium coffee pod. Leaving aside the issues of aluminium manufacture (dirty mining, huge energy consumption), the pods present a complex recycling issue. Some are plastic with aluminium lids that can't be put in domestic recycling, so have to be separated before the plastic can be recycled. Some are aluminium, and can be recycled through brand-led drop-off or courier-bag return programmes — but you have to wash and dry the pods, and keep the bag (or branded box) hanging around your kitchen until you've collected enough to make the trip to the drop-off point worthwhile.

Do you really have the energy for all that? (Maybe that's why you feel like you need a coffee.) If not, there are now 'ethical' coffee pods on the market which can be composted, breaking down within 180 days. Again — you actually have to do this — throwing them in your domestic waste doesn't have the same effect. Or better yet, go old-school and make the change back to a plunger or stovetop brew.

Compost your scraps: Composting is one of the most beneficial ways to recycle in your home, giving you a first-hand experience of transforming waste into useable nutrients and connecting you directly to the cycles of nature. Accurate statistics on household composting are hard to find, with reported numbers varying from 7 per cent of people 'actively composting' (as opposed to chucking veggie scraps in a pile in the backyard) to as many as 60 per cent doing it 'a lot of the time'. For more tips on composting, see Chapter 14: The Landscape.

CHAPTER 9

THE LIVING SPACES

A built-in seating nook at the Auckland home of architects Lisa Day and Scott Donnell uses space that is often wasted. Photo: Duncan Innes

When you imagine your dream home, the living areas are usually the centrepiece of your vision. You picture yourself relaxing in front of a TV show after a hard week at work, sitting around the dining table with friends or family for a meal, or lying on the floor playing a board game with your kids. You imagine ranch sliders or bifold doors thrown open to the sunny garden, deck or patio in summer. You see yourself curled up comfortably on the sofa with a book or iPad on a cold winter evening. You mentally arrange your art, books, musical instruments, stereo equipment and treasured family mementos. You consider how you might transform it into a temporary office or place for overnight guests to stay.

The living space has to play a wide range of roles. It's usually the first area that you bring visitors into, so naturally you want it to be beautiful, welcoming and comfortable. At the same time, because it's heavily used, it needs to be practical, easy to keep clean and durable. The design decisions made here often set the tone for the overall interior design of the home, so it's worth considering how the materials, lighting and furniture style you choose might flow through to other rooms.

Once a series of formal rooms used for entertaining guests away from the rest of the house, today living spaces are informal and multi-functional. Modern layouts are usually open-plan or 'broken-plan' (separated into different zones through the use of stepped floor levels, room dividers, or arranged around a central courtyard). Many owners of villas, bungalows and Art Deco houses choose to open up and extend the traditionally small, separate living rooms to modernise, add light and floor space, and improve the indoor–outdoor flow.

From an eco-perspective, since it's an area where you spend a lot of time, it's essential that it's easy to keep warm in winter and cool in summer, and that the air quality is excellent. The living space usually acts as a transition zone between indoors and outdoors, taking advantage of the best views with large windows, so it makes a big contribution to how connected to nature the house feels. The furniture you buy for the living spaces will be investment pieces that are expected to last a long time. And since the living area is also usually the largest zone in the house, it accounts for a big chunk of the heating and lighting bill.

WALL AND CEILING TREATMENTS

The living and dining areas collectively make up the largest continuous space in the house, so your choice of material or covering for the walls and ceilings will have a big visual impact — whether you want them to take a quiet back seat to artwork or furnishings, or to make a statement of their own through colour, pattern or texture.

If you've chosen a structural material that also acts as an interior surface finish, like tilt-slab concrete, rammed earth, exposed brick or stone, or structural timber panels, the decision is already made. But if you're working with plasterboard, or are looking to reline, your choice of wall treatments can make a significant difference to the air quality within your home, as well as the kinesthetic feel of the space.

PLASTERBOARD

Plasterboard or drywall, best known in New Zealand as GIB or Gibraltar board (so named to suggest that it's as dependable as the Rock of Gibraltar), is the most common interior finish. Plasterboard is made from compressed gypsum plaster sandwiched between two pieces of cardboard, usually made from recycled paper. Gypsum is a naturally occurring mineral, calcium sulphate dihydrate, which is mined in Australia and shipped to New Zealand to be made into plasterboard by Winstone Wallboards. This is an energy-intensive process that produces greenhouse gases. So although plasterboard that doesn't contain synthetic additives or fibres is a non-toxic product in itself, it has a high embodied energy cost, particularly plasterboard with no recycled content. Considering that 100 billion square feet of plasterboard is produced around the world every year, it's a building material that makes a significant contribution to climate change.

Clean-waste plasterboard left over from building is highly recyclable, so any offcuts or waste should be returned to the manufacturer for recycling into more plasterboard, composting or even being converted into an agricultural soil-conditioning product. Even better is careful planning to minimise waste in the first place. Plasterboard that has been painted with environmentally friendly paints may also be able to be recycled. Conventionally painted or papered plasterboards probably can't be, however, and have to be disposed of in landfill. As much as 9 per cent of landfill waste is thought to be waste plasterboard.

There is some imported plasterboard available in New Zealand, although the vast majority is manufactured here. After Hurricane Katrina in the United States, issues arose with some imported plasterboards manufactured in China, which were shown to release high levels of corrosive and irritating hydrogen sulfide gas. Although there don't seem to have been any similar issues here, locally made plasterboard is certainly a better environmental choice for its lower carbon kilometres. Specifying 13 mm-thick plasterboard (instead of the standard 10 mm) will add insulation value to walls, particularly internal walls where plasterboard is used on both sides.

NATURAL PLASTER

Fans of natural plasters tend to rave about the unique feel that they give to a home — a kind of softness and gentle aesthetic variability that you don't get from paints, wallpapers or hard surfaces like stone, brick or concrete. You can paint over plasters, though they do contain natural pigment, which

'Second-grade' ply — that's plywood that's been classed as 'inferior' due to knots or finish — was sourced for the panelling of the Gittos House. Sometimes the difference between second- and first-grade ply is hard for anyone but an expert to tell. If you have the time to go through a pile choosing pieces, it's a good way to save money. Photos David Straight

can be enhanced with low-VOC colourants. Unpainted plasters should be sealed with an oil or wax-based sealant to keep dust down, reduce moisture absorption and give them longevity.

There are a range of options available in New Zealand for natural plaster interior finishes, ranging from traditional lime plasters, clay plasters, gypsum plaster to polished plaster that contains powdered marble, although finding plasterers who work with the more specialist materials can be difficult. It's worth the effort to find one though, as although it's possible to acquire and mix the materials for lime or earthen plasters yourself, achieving clean, smooth surface results can be very difficult, so it's best done by or with the close advice of a professional.

Acrylic plasters contain petroleum products which can off-gas, so are considered less environmentally friendly. Cement plasters, while long-lasting, have a higher embodied energy cost than gypsum plaster.

WOOD PANELS

Wood brings a soft warmth and beautiful natural ambience to a home, and has been particularly popular as an interior finish in New Zealand since the middle of last century. There are many ways of using hardwoods or softwoods as interior claddings, from sustainably sourced new or recycled tongue-and-groove planks, to sheet panels, either solid, veneers or manufactured plywood or fibreboard. As covered in Chapter 6: The Structure, wood is an eco-friendly building material when it is sustainably grown, thanks to its renewability and carbon-sequestering qualities.

Although native timbers were extensively used for early New Zealand home-building, there are few native forestry timbers now available for interior wood panelling, and all are grown in privately managed forests. There are also good supplies of recycled native timbers available through demolition yards and specialist suppliers.

A number of sustainable exotic plantation timbers are grown in New Zealand. Macrocarpa, which is often used as an all-purpose term for cypresses, is one of the most popular interior timbers for its warm golden-brown tone and natural resistance to borer. Lawson's cypress, a paler cypress, is also popular for a more Scandinavian effect. There are also a huge number of imported wood options, such as Western red cedar, American oak and ash, jarrah, walnut and many more. These can usually be sourced as sustainable plantation timber with well-recognised certifications, though they have the obvious drawback of needing to be transported long distances to get here.

Wood panels come in several types — plywood, which is made of very thin sheets of wood glued together to give the sheet strength and warp-resistance, and medium-density fibreboard (MDF), particle board and oriented strand board (OSB), which are made from sawmill waste held together by resins. MDF contains the smallest wood particles, and OSB the largest. Recycling sawmill waste is a positive for the environment; however, the main drawback for these products is that the resins used in them usually contain formaldehyde, a known carcinogen that can off-gas into the environment.

There are some formaldehyde-free products, but most panels carry an emissions rating for the amount of formaldehyde that they release. Ask for panels with a rating of E0 or E1, which are the highest standards for indoor air. As always with wood products, check to ensure that the product you're specifying carries a sustainable forestry certification. Some products also carry Environmental Choice certifications, but just because one product in a range carries it, it doesn't mean all will, so check every individual product. When sealing wood panels, choose natural finishes that use plant-based oils, tree resins and waxes, in order to avoid off-gassing.

PAINT

Since the days of cave-painting, paints have undergone a huge evolution. Early paints used earth, plant and mineral-derived pigments mixed with natural gums and binders (such as egg yolk or milk) and thinned with water and oils. Of course, just because an ingredient is natural, it doesn't mean it is safe — mercury, arsenic and lead have all been used as paint colourants in the past. See Chapter 7: The Shell for advice on dealing with lead-based paint.

Modern commercial paints — and associated products, such as primers and varnishes — are highly synthetic, often containing a complex mix of petrochemicals, heavy metals, fungicides, formaldehyde, polyurethane and dozens of other chemicals. They are typically made up of solvents (liquids), resins (binders for adhesion and durability), colour pigments (especially titanium dioxide and zinc oxide) and other chemical additives that require large amounts of processing. The production process for conventional paint is hugely wasteful, consuming large amounts of energy and fossil fuels, and creating toxic waste products that are difficult to dispose of safely.

People often talk fondly about that 'new house smell'. Unfortunately, that smell is really a heady mix of volatile organic compounds (VOCs) — gases that have potential to cause short-and long-term health effects, from headaches, nausea, allergies and skin irritation, to an increase in certain types of cancer. And even after the smell disappears, paints can continue to emit VOCs into the air for several years. When paints containing VOCs are used inside, it can create significant air-quality issues.

Most of New Zealand's major paint companies, such as Resene and Dulux, now offer a range of water-based low-VOC and VOC-free paint formulations. These are considerably better for indoor air quality, although they may still contain many synthetic ingredients and petrochemicals. These paints are harmful to the environment if disposed of incorrectly, so never pour them down your sink or into a drain. Instead, return leftover paint to the manufacturer for correct disposal, check to see whether your local council has a hazardous waste disposal programme, or search for a local chemical waste management company to take care of it for you.

Modern natural paints have made huge advances in the last couple of decades. They're now available in a beautiful range of shades and finishes, and if applied correctly, will last as long as synthetic paints. Better yet, the manufacturing and disposal processes are considerably better for the environment. As you would expect, there's a price premium for natural paints, sometimes as much as twice the cost of standard paints. New Zealand's largest natural paint company, the Christchurch-based Natural Paint Company, uses china clay, chalk, waxes, plant oils and tree resins to create their interior and exterior paint products.

Recycled paints are available in New Zealand through the company EnviroPaint. These paints represent a reduction in both the environmental impact of paint production, and a reduction in waste paint (and associated products like tins and pails) being sent to landfill. The range of colours available is much more limited than you will find in a new paint range, but the price is very comparable to new paint (and sometimes lower), making it a good eco-solution for those on a budget. Make sure to look for recycled paint with low-VOC levels.

WALLPAPER

Wallpaper might seem like a natural eco-choice, since it's made from a renewable resource, but the glues and vinyl coatings used on some papers can contribute to poor air quality in the home, and prevent moisture from 'breathing' through plasterboard freely.

There is an extensive range of beautiful patterned and textured wallpapers available in New Zealand, some manufactured here, and many imported. Some wallpapers are literally made from paper (ask for paper that is sustainably forested, or contains recycled content) and cellulose, but in other cases, the papers are made of PVC (polyvinyl chloride), or have a PVC backing or coating to make them more durable or able to stand up to wet cleaning. PVC prevents moisture from moving through walls, which can cause mould build-up. To combat this, wallpapers may contain fungicides.

Other types of eco-friendly specialist papers exist — fibreglass wallpapers, which use woven glass fibres, quartz and lime to create a strong and sometimes textured wall covering, bamboo wallpaper, and grasscloth, which is woven with natural reed and jute fibres.

When choosing wallpapers, ask your wallpaper retailer to recommend a water-based adhesive with low- or zero-VOC levels if this is suitable for the type of wallpaper you've chosen, and whether the wallpapers have been printed with environmentally friendly latex-based or water-based inks.

One more note. If you're searching online for 'eco wallpaper', you'll probably be delighted to discover the beautiful Scandinavian brand Eco Wallpaper as one of your first hits. However, be aware that the word 'Eco' is actually a contraction of the company's original name, Engblad & Co, and is not a specific reference to environmental friendliness, although the company does claim to have a commitment to looking after the environment. At any rate, the papers have to travel a long way to get to New Zealand, so carbon kilometres need to be a consideration with this brand.

LIGHTING

For many decades, the main choice in home lighting was the incandescent (or tungsten) lightbulb — cheap, widely available and with a pleasant warm glow, but also hot to touch, short-lived and terribly energy inefficient, converting only 5–10 per cent of the energy that it uses into light.

Today, there are more eco-friendly choices available: compact fluorescent lamps (CFLs), halogen lamps, and LED (light-emitting diode) bulbs. Of these three, LEDs are the most efficient, and have taken giant strides forward in performance since they first came onto the market in the mid-2000s, with the cost also coming down substantially. They can be used for every form of lighting — pendants, spotlights, recessed lights, lamps and wall lights.

Around the world, many countries, including Australia, the United States, the European Union, Canada, China, India, Malaysia and South Africa have

SHINING A LIGHT ON CLIMATE CHANGE

It's not just your bank balance that benefits when you change to an energy-efficient lightbulb, it's the climate. According to the New Zealand Electricity Authority, lighting accounts for up to 12 per cent of residential electricity use. When the Australian government announced its decision to phase out incandescent lightbulbs in 2007, Malcolm Turnbull (then the Minister for the Environment) estimated it would save 8000 tonnes in greenhouse gas emissions a year. Australia has much 'dirtier' power than New Zealand (relying on coal for most of its generation), so the emissions savings there would be much bigger than what we would make with a full phase-out. But given that New Zealand has the fifth highest greenhouse emissions per capita in the OECD, every little bit counts. And when it's as simple as replacing a lightbulb — and then forgetting about it for the next 15 years — this is one of the easiest eco-choices to make.

Open treads and vertical rod dividers allow natural light to pass into a stairwell at the Gittos House. Photo: David Straight

implemented full or partial phase-outs of incandescent bulbs. Here, however, the government has left it as a matter of 'consumer choice'. Whether due to concerns about price, misconceptions about performance or a lack of knowledge about efficient lightbulbs, consumers aren't switching over of their own accord, with less than a third of lightbulbs sold through supermarkets (where most people buy them) being 'energy-efficient' ones.

And that's surprising, because there are big savings to be made over the long term by changing to energy-efficient bulbs. In the case of replacing a 60-watt incandescent with a 9.5-watt LED (which gives out the same amount of light), it takes just a little over a year for the higher upfront cost of the LED to be covered by savings in electricity costs.

After that, the LED represents average savings of $15 a year (including the cost of replacing the incandescent bulb every year). With LED bulbs lasting up to 15 years, that adds up to a couple of hundred dollars in lifetime cost savings for every incandescent bulb replaced. Although local figures aren't available, the US government estimates that the average American home uses 40 lightbulbs, and it's likely to be similar here. That's $8000 in lifetime cost savings for a full switchover. How many accountants does it take to change a lightbulb again?

Of course, there's one kind of lighting that's even more eco-friendly than LEDs — and that's natural light. Allowing more daylight into your home reduces the need for artificial lighting, adds warmth through passive solar gain, and has been shown to bring health and psychological benefits through increased vitamin D levels and improved circadian rhythms. Finding ways to let more natural light into your home is a great first step in reducing your energy use and improving your sense of well-being.

DAYLIGHTING

The living areas, kitchen and working areas (such as home offices) are the parts of a home that benefit most from receiving lots of natural light, though introducing daylight into traditionally darker areas like hallways and stairwells can have an uplifting effect and make a home feel more expansive.

Many houses built before the 1970s in New Zealand have small windows in comparison to more modern builds. Enlarging existing windows lets more natural light in, and can be a good opportunity to improve the thermal performance and airtightness of the windows. Raising the height of a window usually lets in more light than widening it does, as it increases the proportion of sky in the view, and the amount of brightness in the home.

Removing internal walls and doors is another way to let light flow more freely through a house. A non-load-bearing wall can be knocked out and refinished over the space of a couple of days for as little as $2000. Don't forget, however, that allowing light to flow more easily also allows heat to flow. Before you pull out the sledgehammer, think about how removing that wall could change the heating requirements of the space.

As an alternative, a glazed door or internal window will let light pass through rooms without losing heat in the same way. Other, more subtle ways to let light move more freely through a space include 'floating' cantilevered staircases and walk-on glass panels in upstairs levels.

Skylights allow large volumes of light to pour in (and give beautiful night-sky views), but care should be taken that they don't add too much solar gain from heat falling directly onto thermal mass, or allow heat to escape at night. Thermally broken frames, sky-blinds, and double- or triple-glazing with low-E glass are important ways to control this kind of heat loss. High-level clerestory windows that allow light to enter the house without adding too much solar gain may be a good alternative to skylights.

For around $1000, a solar tube is an easy way to bring natural light into a dark area, particularly through a pitched roof. A roof dome introduces light through a reflective light pipe, which is then diffused through a ceiling fixture. Some solar tubes can be integrated with ventilation systems for bathrooms, solving two problems at once.

ENERGY-EFFICIENT LIGHTBULBS

When replacing your traditional incandescent bulbs, there are three main alternatives. Be aware that some bulbs boldly marketed as 'eco' on the packages are not actually the most efficient ones. Read the description carefully to see what you're getting.

In Strachan Group Architects founder Dave Strachan's own home, he employed the innovative technique of using elevated concrete slabs as thermal mass below high windows to capture winter sun that otherwise wouldn't reach the concrete floor slab. Photo: Simon Devitt

Okay: halogen

Cost: $2+

Halogen bulbs, often used in ceiling downlights, are incandescent bulbs that are somewhat more energy-efficient than standard bulbs (around 25–30 per cent). They last around two to three years compared to one year for traditional incandescents, and are still very hot to touch. If you have halogen bulbs in your ceiling downlights, it is a fire hazard to have insulation over the top of them, meaning your home may be dotted with little 'chimneys', where heat can be lost through the uninsulated light fittings.

Better: compact florescent lamps (CFLs)

Cost: $5+

CFLs are scaled-down versions of the long florescent light tubes that we're used to seeing flickering in offices and schools. CFLs use about 75–80 per cent less energy than traditional incandescents, and last at least six years. Drawbacks include the fact that they need to warm up for a minute or two before they reach full brightness, that they need good ventilation to work, that many are not dimmable, and also that they contain a small amount of mercury, which is an environmental contaminant. Don't use them in areas where they may be easily broken (particularly children's rooms). If one does get broken, you'll need to vacate and ventilate the room for 15 minutes before following the manufacturer's clean-up instructions. Used CFLs should be placed in a plastic container or glass jar for disposal in your outside bin, or ask your local CFL retailer for recycling options.

Best: light-emitting diode (LED)

Cost: $8+

LEDs are now the gold standard in energy-efficient lighting. They come in a wide range of luminosities and colour tones, from warm/yellow (this will be written on the box as something like 2500 K) to cool/blue (up to 5000 K). They are much cooler to touch, meaning ceiling insulation can be laid over the top of some types of LED ceiling downlight, helping to create a complete thermal envelope. They are 80 per cent more efficient than standard incandescents, and they last at least 10 times as long, with some manufacturers claiming a lifetime of up to 50,000 hours (50 times as long as a standard incandescent lightbulb).

DID YOU KNOW?

Modern dimmer switches save power when you're using them with LED bulbs, but are actually less energy-efficient when used with traditional incandescent bulbs.

HEATING

With space heating accounting for around 30 per cent of the average Kiwi home's power bill, and the living areas using a large portion of that, it's an important place to get it right. The World Health Organization recommends that living areas be kept at an average temperature of 21 degrees Celsius year-round. This means almost all Kiwi homes need heating at least in the evenings for several months of the year.

According to *Consumer* magazine, 38 per cent of Kiwis' homes are not as warm as they would like them to be, and a 2013 Massey University study showed that a small percentage of homes (around 3 per cent) have no heating at all. As nice as it would be to believe that's because they're so energy-efficient they don't need additional heating, it's more likely that these households simply cannot afford electricity. And electricity is what 86 per cent of Kiwis use to heat their homes. Considering the poor insulation levels in many homes, that means a good chunk of many power bills is wasted heating. If your home is hard to heat and poorly insulated, addressing the insulation issue is more important than investing in a more powerful heat source. Wherever possible, look for ways to use passive strategies like thermal mass to gather and store heat naturally through sunlight.

FIRES AND WOOD-BURNERS

It's hard to beat the pleasure of warming yourself in front of a dancing fire. Flames have a mesmerising ambience we're naturally drawn to — nobody daydreams about snuggling up in front of a roaring heat pump on a cold evening. Fire's elemental danger and unpredictability are exciting, but when under control the warmth and quiet pop and crackle of a fire is so soothing that it's been proven to lower blood pressure. From a biophilic perspective, fire is the only home heat source outside of sunshine that gives us a direct sense of connection to nature.

Open fireplaces were a typical feature of New Zealand villas and bungalows, and although many have been closed over, some still remain, and are used as a central source of heating. As an eco-option, open fires — even the fancy designer ones — have several drawbacks. They're terribly energy inefficient, as most of the heat is lost up the chimney (which continues to siphon warm air out of the house when the fire goes out), and they pose a fire-danger risk from sparks. Smoke escaping into the living area also diminishes indoor air quality, and since open fires produce 'dirtier' smoke than modern wood-burners, they create worse air pollution.

If you have an existing open fireplace that works well for the layout of your room and can afford to convert it to an inset wood-burner, this can be a beautiful, energy- and cost-efficient way to heat your living spaces, since the masonry of the fireplace will act as thermal mass to store and slowly release heat even after the fire is out. And if you can access a free or cheap source of

Left: A wood-burner supplies all the heat needs of the living space at architect Lisa Day's own eco home. The brick setting helps store heat. Photo: Duncan Innes

Below: Firewood is stacked for seasoning under a covered passageway at the K Valley House. A powerful wetback wood-burner keeps the open-plan living space warm and water heated through the winter. Photo: Patrick Reynolds

firewood like local treefall, your running costs will be very low. Even if you have to purchase firewood, the costs compare closely to efficient heat pumps.

Modern freestanding wood- and pellet-burners (where the pellets are made from compressed sawmill waste) are far more energy-efficient than open fires, and some can be constructed with a 'wetback' to heat water. Wetbacks obviously use some of the energy of the fire that would otherwise be used to heat the house, so make sure the burner is big enough to adequately heat both your home and the water. If you have a modern wood-burner, you'll know that they can generate a large amount of heat, so investing in a heat transfer system can mean one big wood-burner can quite easily heat an entire house. And, of course, fires have the old-school benefit of working even when there's a power cut.

Burning wood from a sustainable forest is carbon-neutral, and although modern wood- and pellet-burners do release particulate matter in the smoke, the emissions are far cleaner than the smoke emitted from open fires and older styles of wood-burners. The main disadvantage of wood-burners is that you can't turn the heat on at the flick of a switch — they take a while for warmth to build up after they've been lit.

Ultra-low-emission down-draught burners are a fairly recent arrival on the market, and take the 'clean fire' concept to a new level. They've been approved for the Canterbury market, which is one of the most stringent in

HOT TIP: MAKING THE MOST OF FIREWOOD

Just like food, firewood is best when it is well seasoned. Seasoning firewood means storing it outside, loosely stacked and covered on top, with some wind and (ideally) sun exposure to dry it out, for at least six months (softwoods) to 18 months (hardwoods). When cracks appear in the ends, it's ready. Most commercial firewood is pre-seasoned.

- Aim for a maximum 15-cm diameter for the most efficient burning.
- Never burn treated or sealed wood, particle board or wood with plastics attached (like MDF), as they release toxic fumes.
- Have your fireplace and chimney professionally cleaned every year to keep it efficient.

the country. Both imported and locally made models are available. One thing to be aware of with these burners is that their surfaces can get extremely hot, so if you have small children, you need protective guards in place.

Many local authorities have restrictions on the kinds of wood-burners that can be installed, so be sure to check that your model is approved before you buy. The Ministry for the Environment has a list of approved wood-burners on its website. You will need a building consent to install, replace or relocate a wood-burner on your property.

Bioethanol burners are often promoted as a clean or 'green' alternative to natural gas or wood-burners as the fuel source comes from plant-derived ethanol (produced through fermentation). However, they're nowhere near as powerful as wood-burners or heat pumps, and even if you could generate the same amount of heat, the cost would be substantially higher. As an attractive accessory heater, they're an eco-friendly option. Make sure the room is well-ventilated, as the burners consume oxygen.

HEAT PUMPS

Heat pumps are an increasingly popular heating option — more than a third of Kiwi homes have them — and in most cases (except for small, well-insulated rooms that don't need heating for long periods of time) they're the most energy-efficient and cost-effective way to use electricity to heat your home. They work quickly, and are generally quiet, though less so than other heating options.

Air-to-air heat pumps are the most common type in New Zealand. They work by extracting warmth from outside air and transferring it indoors. They can work the opposite way, too, acting as air conditioning to cool the house in summer. Heat pumps are usually designed to heat a single space, but can be installed with several outlets or a ducting system to heat several rooms.

Although heat pumps are much more efficient at converting energy to heat than electric heaters, they won't necessarily decrease your electricity use. Don't make the mistake of thinking you can use a single-room heat pump to heat an entire house just by leaving internal doors open. You'll overwork your appliance and send your power bill skyrocketing. And heat pumps have to be sized correctly — too small, and they'll run all the time trying to keep the space they're heating at the set temperature; too big and they'll flick on and off constantly, reducing efficiency and wearing the unit out.

Their filters have to be cleaned often (monthly, or even twice a month if you use them heavily) to remove dust, dirt and animal hair, or they become less effective. If you live in a particularly cold part of the country, where temperatures drop below 6–7 degrees, you may find that heat pumps are not as efficient as other forms of heating. And if you're not currently using electricity for air conditioning in summer, installing a heat pump and using it for both heating and cooling can result in a larger overall power bill for the year.

Although heat pumps blow air into the room, it is recirculated air, not fresh air, so they don't have any ventilation benefits. If you're putting a heat pump into your open-plan living space, it shouldn't be installed near the kitchen, as they can end up drawing in fat-laden steam, which will eventually start to moulder and smell when recirculated.

HEATERS

Electric and gas heaters are best as backup options to a bigger, more efficient heat source like a wood-burner or heat pump, or for small rooms with short-term heating needs. For a big space like a living area, they're less ideal.

Electric heaters come in several different types, ranging from tower fan and upright oil heaters, to micathermic, infrared and panel and convection heaters. Oil heaters are generally more power-hungry and slower to heat the room than ceramic tower fans or micathermic heaters.

Flued gas heaters and gas fireplaces have become a less popular heating option as more people have turned to heat pumps, even though sleek modern gas fireplaces have come a long way since the days of gaudy faux-wood logs. They burn piped gas — which is available only in the North Island — or LPG as a heating fuel, with the emissions being released through an out-pipe. Gas is a non-renewable resource, and New Zealand's fields are currently declining, which is likely why gas prices have been rising faster than electricity prices. Depending on the price you pay for gas, this might still be a more cost-effective option than electricity — at least for now. Portable unflued gas heaters release moisture and hazardous gases like nitrogen dioxide and carbon monoxide. They're widely considered the worst and most dangerous form of indoor heating, and should be avoided.

RADIANT HEATING SYSTEMS

Radiant heating systems produce heat by passing warmed water through floor, wall or even ceiling pipe. Because of the high cost and the difficulty of retrofitting them, these systems are usually installed in new builds. They are popular overseas but haven't caught on much in New Zealand yet, which is a pity, as they can be a good choice, especially in cooler parts of the country.

Underfloor radiant systems work best with high thermal mass materials like concrete, tiles, stone and even wood. The material should be exposed for the best results — putting carpet over an underfloor heating system will prevent the heat from radiating effectively. Depending on the type of living space you have (for example, high-ceilinged with lots of glass), you may need a secondary source of heating in addition to an underfloor system to achieve a comfortable temperature.

Radiant heating systems are quiet, don't circulate dust easily, and are built to last for a very long time (50+ years for the pipes). Some people find it hard to get their heads around the idea of heating a whole house rather than just the rooms that they're using. This is a peculiarly Kiwi approach to heating, one that's not shared by most of the rest of the world.

FIVE RULES FOR CHOOSING SUSTAINABLE FURNISHINGS

1. Choose recycled

Back in our grandparents' day, furniture was really built to last — and much of it has. There are many wonderful, characterful pieces of antique and mid-century designer furniture in excellent condition available from second-hand furniture retailers and opportunity shops for reasonable prices. Beware of second-hand replicas of well-known designer pieces — these are not manufactured to the same standards as licensed pieces. Look for tags or stickers that show the licensed manufacturer's name, or familiarise yourself with the key details of the originals online.

Well-loved but solid pieces of recycled furniture can be given a facelift with a new coat of paint, just make sure you use zero-VOC or low-VOC paints, and let the piece stand outside for a couple of weeks to off-gas before moving it into the house. Fresh upholstery is another option. Tracking the supply chain and chemical content of these products is obviously difficult, but the good news is that if they did use finishes that contain VOCs, they will have off-gassed completely many years ago.

Some fabric manufacturers are now producing upholstery fabrics made from 100 per cent recycled post-consumer waste, such as soft-drink and water bottles. This is a great reuse of materials that usually only have a short lifespan, but thought should also be given to where this fabric will go at the end of its life. Ask the retailer or manufacturer for their advice.

2. Choose quality

There's no question that high-quality furniture is expensive — as it always has been. This can be hard to get our heads around, given that over the last 40 years, the price of many household items like televisions, appliances and stereo equipment has dropped while the quality has improved. This is because the component prices and manufacturing costs of technology have plummeted, but the costs of materials like wood, leather, metals — and the cost of the labour to assemble them into high-quality furniture — have remained relatively constant. But that's a reflection of the longevity of the product. While your iPod speaker system may be out of date (or mysteriously failing to function) within five years, a well-made lounge suite or table can last decades, and be passed on to the next generation.

3. Choose multi-functional or built-in

If a piece of furniture can do more than one job, that's another piece of furniture that doesn't need to be produced, bought and disposed of. Wall beds, fold-down tables and chairs, sofas that convert to a guest bed, built-in seating with a bookcase at the back and built-in shelving all save space and material costs.

4. Choose natural materials

Where possible, choose solid natural materials that have been sealed or treated with natural stains and finishes, rather than composite products like MDF. Polyurethane (also frequently used as a flooring finish) is a furniture varnish that can emit VOCs including formaldehyde for many years. The Eco Design Advisor service recommends linseed oil, tung oil and beeswax as alternative finishes.

Although they don't meet the approval of animal welfare organisations like PETA, wool (particularly felted wool and mohair) and leather are durable, hypoallergenic and naturally stain-resistant, making them good choices for seating fabrics instead of vinyl or other synthetics. Organic linen, velvet, hemp cotton and thick calico (which is low-processed unbleached cotton) are good plant-sourced choices, though they're generally less durable and stain-resistant. Latex foam, and wool and cotton wadding are good stuffing choices.

As always, look for certifications that the products have been sustainably produced, such as FSC-certified timber and the EnviroChoice stamp.

5. Choose recyclable

If budget constraints mean you are limited to mass-market furniture brands, make sure to ask about their recycling programmes. Overseas, IKEA has started a pilot programme for returning used furniture at the end-of-life. Consumers, too, have started a repurposing movement known as 'IKEA hacking', where standard pieces are customised or transformed to have new functions.

Good-quality used furniture can be donated to charity shops, or put out in the inorganic rubbish collections, where it will also be recycled if possible.

GEOTHERMAL HEATING

The word 'geothermal' can give the impression that these systems tap into bubbling underground hot springs. In fact, the name means that they draw heat (thermal) from the earth (geo). They're also known as ground-sourced heat pumps. Underground temperatures are usually higher than air temperatures, particularly in winter, and are stable year-round. Geothermal systems use a network of pipes filled with fluid buried underground to extract heat from the earth and turn it into hot water that is circulated through an underfloor heating or wall-based radiator system.

The systems are more energy-efficient than air-sourced heat pumps, but they're expensive to invest in upfront — typically more than $25,000. This is the sort of system to consider if you're planning to stay in your house for a long time.

FURNISHINGS

The furniture, art, decorative pieces, books, plants and other personal items that we put in our living spaces are what really make them live. But many common household items contain VOCs that can off-gas for years. The chemicals known as halogenated flame retardants that are used on soft furnishings like sofas to increase their fire-resistance, and others used to increase stain-resistance on rugs, have also been flagged as concerns by green building watchdogs like BuildingGreen.

Cheap, fashionable, mass-market furniture is made to last just a few years before breaking or being discarded in favour of a 'new look' — and most of it goes into landfill. In Australia, the average household throws away 24 kg of wooden furniture per annum — if figures are similar here, that means almost 40,000 tonnes of it go to landfill every year, rising to an estimated 51,000 tonnes by 2038 (and that's if the amount we throw away doesn't increase). That's not even counting plastic and metal furniture. Looking at furniture as a lifetime investment is an important way to reduce your household waste footprint.

BRINGING THE OUTDOORS IN

House plants have seen a recent resurgence in popularity in interior design magazines, but they're not just an aesthetic feature. Plants absorb carbon dioxide and release oxygen. Green walls, also known as vertical gardens, can be a striking interior feature, but they can also be hard work to keep alive. A mix of big, large-leafed potted plants and smaller leafed trailing plants and ferns can give a similarly lush effect with lower maintenance. Large terrariums, especially ones that include succulents, are extremely low-maintenance (their biggest problem is usually over-watering) and are ideal for those not blessed with a green thumb.

The Wairarapa Haybarn by First Light Studio uses an underfloor heating system. Photos: Jason Mann

CHAPTER 10

THE BATHROOM

Glass doors separate the shower area from the rest of the bathroom at this Titirangi home. Photo: Simon Devitt

In a country so frequently rain-lashed as our Land of the Long White Cloud, it can be easy to forget that water is a precious resource in need of conservation. That attitude has led to New Zealanders consuming more water per capita than many other developed nations. With many regions experiencing regular droughts that are likely to become more severe with climate change according to the Ministry for the Environment, water consumption is an issue we can't afford to be casual about.

The average New Zealander uses around 150 litres of water a day indoors — or a little over 54,000 litres a year — almost a fifth of which is flushed straight down the toilet. As much as 10 per cent may be dripping away as undiscovered leaks. Multiply that amount by our population, and you are looking at over 250 billion litres of water consumed by Kiwi households every year.

Bathroom use accounts for about 70 per cent of our indoor water consumption. And although all our tap water is treated to drinking quality (using energy and resources) we only drink around 2 per cent of it. Our bathrooms are also heavy users of electricity — water heating uses about a third of a home's energy. This means there is a huge opportunity to decrease the amount of treated water we consume by using tank rainwater or greywater (in certain areas only) for non-drinking uses like toilet-flushing, and decrease our power bill significantly by investing in better water-heating systems.

WATER HEATING

Waikato-born electrical engineer Lloyd Mandeno invented one of the world's first electric water heaters back in 1915, and patented a version eight years later. Since then the hot-water cylinder has become the first choice of more than 80 per cent of New Zealanders for water heating, despite not being the most efficient one. Gas is the second most popular option, and can be more energy-efficient than an electric system, though not necessarily cheaper unless you use gas for multiple services in your home. Gas is the least eco-friendly heating option you can choose, as it's a non-renewable resource.

If you have a hot-water cylinder, particularly one that's more than 10 years old, it is easy and worthwhile to wrap the cylinder with an insulating wrap to decrease its continual 'standing loss' of heat, and to 'lag' at least the first metre of the cylinder pipes with an insulating foam tube. If your home is on a raised foundation, you may be able to find hot-water pipes running underneath the house to the kitchen to lag as well. Gas hot-water systems shouldn't be wrapped because they require ventilation, but their pipes can be lagged.

Heat-pump hot-water systems draw energy from outside air in the same way as space-heating heat pumps. They are more efficient than electric hot-water systems, particularly in parts of the country where temperatures don't often get below 6 degrees Celsius. Although the cost of purchasing and installing one can run to $5000, they can reduce your water-heating costs by two-thirds. As the units can be noisy, make sure they're placed away from bedrooms and neighbours. The Ministry for Business, Innovation and Enterprise recommends researching your choice carefully, as some units are much better than others.

Solar hot-water systems (boosted by electricity, gas or paired with wetback heating) are worth the investment as long as your home receives enough sun to heat the system and the installation is done well. Done right, you can expect one system to heat up to 75 per cent of your hot water for free, making them the cheapest to run (although they can be difficult to retrofit, so usually work best in a new build). Choose a system where the cylinder is placed inside the house rather than on the roof, if possible, especially in colder regions. Showering or bathing in the evenings is best with these systems.

Of course, another way to reduce your hot-water bill is to reduce your water use. Taking shorter showers (or turning the shower off while shampooing or lathering up), fixing dripping taps, turning the tap off while you brush your teeth, showering instead of bathing, and putting the plug in when you wash your hands are all small actions that accumulate big differences over time. You can also invest in built-in gadgets and systems that do some of the water saving for you.

BASINS

A gushing tap can send 15 litres of water a minute down the plughole. And the design trend towards basin mixers that run hot and cold water simultaneously means you may be bumping your power bill up every time you wash your hands — even when the hot water doesn't even make it to the basin by the time you have finished rinsing. Choosing separate hot and cold taps makes washing with cold water only easier. But if you're fixed on a mixer, make it a habit to swing the handle all the way over to the right and leave it there afterwards.

To check the flow rate of your tap, run the water on full blast into a 2-litre-plus measuring jug for 15 seconds. Afterwards, multiply the amount in the jug by four to find your flow rate. Around 6 litres a minute is what you should aim for — if it's much higher than that, consider fitting a flow restrictor (available for less than $20 from plumbing and bathroom suppliers) or upgrading to lower-flow taps. Look for basin taps with a WELS rating of 4 stars or higher. Some modern home faucets come with sensors that activate automatically when a hand is waved in front of them, and shut off when the hand moves away.

The Rammed Earth House by Assembly Architects uses a continuous tiled surface on the deep, Japanese-style tub and an open shower that can be used with a stool for soaping up pre-bath. The light from the left comes in from a tall, openable window for natural ventilation. Photo: Simon Devitt

HOW TO SOAK JAPANESE-STYLE

In Japan (and some Scandinavian cultures), taking a long soak in a hot bath is an essential practice for relaxation, regeneration and healing. At home, the soaking is done in a deep, short tub called an *ofuro*. *Ofuro* are usually made of a fragrant wood like hinoki (a cypress) or cedar, but can also be tiled or made of copper. Although the depth of the tub means they do use quite a lot of water, several members of a family may soak in the same tub in succession. That's why cleansing yourself first with a *kekeyu* — a soap and rinse in the shower — is part of the tradition.

Soaking in warm water with our eyes closed can return us to something of a primal state that is deeply soothing. It also relaxes muscles, moisturises the nasal passages and skin, and improves circulation. After about 15 minutes, the skin on your fingers will start to 'prune' and your skin can start to become dehydrated; it's time to give it up for the next person.

TIME TO DITCH THE HEATED TOWEL RAIL?

It makes your towels cosy in winter, but running a heated towel rail 24/365 costs you over $150 a year. Changing to thin-but-absorbent Turkish (or hammam) towels and hanging them on hooks instead of a rail could be the answer. If you really can't forgo your warm towels, fitting a timer switch to the heated rail and running it for just a few hours a day will cut your power bill down.

BATHS

In most homes (except those with smaller children), baths tend to be an occasional indulgence rather than a daily ritual. Part of that is time constraints, and part of it is because we've had the message hammered home that baths use much more water than showers. However, that is only true if you're taking quick showers with a low-flow showerhead. An average bath uses 80 litres of water, while a five-minute shower using a 3-star rated low-flow head might use as little as 45 litres. But if you're in the habit of taking 10-minute showers, even with a low-flow head you'll use more water than a bath (which you could soak happily in for much longer than 10 minutes). If you're still using a high-flow head, your five-minute shower could be just as thirsty as a bath.

The takeaway here is that if you're treating yourself to a long soak, indulge in a bath with a clear conscience. But try to balance it by making your daily shower a bit more military in length.

SHOWERS

Showering accounts for about 27 per cent of indoor home water use. Modern low-flow showerheads can reduce the amount of water flow needed for an enjoyable shower from 12 litres a minute (standard fittings) to 6 litres a minute or less. What's even better is that every litre per minute of water that you save can carve over $30 a year off your hot-water bill. If you're on metered water, that bill will go down, too. There are a wide range of low-flow head styles available today. The style you choose will come down to individual preference for aesthetic and droplet feel. Be aware that finer water droplets cool more quickly, so might lead to you bumping up the hot water in the mix in your shower. Look for 3-star WELS-rated low-flow heads.

WELCOME TO THE PLEASUREDOME

Another Kiwi invention, the shower dome, is a clear acrylic lid for your box shower that holds steam in, reducing bathroom humidity and misted mirrors, and keeping your shower warmer, more enjoyable in winter and — surprisingly — clearer. Although ventilating your bathroom after every shower is still a good idea, a shower dome can cut down on bathroom steam considerably.

VENTILATION

Showers (and, to a lesser degree, baths) create steam, and when steam condenses on walls and ceiling surfaces and is left to sit, it's a recipe for fast mould growth. Every shower releases as much as 1.5 litres of water into the air in the form of steam, and if you walk out of the bathroom afterwards leaving the door open, that moisture can disperse straight into hallways and bedrooms nearby. So the bathroom is the most important room to ventilate well — but many of us are not doing so. The BRANZ 2015 House Condition Survey showed that only half of the 560 surveyed homes had bathroom extraction fans, and many of the ones that had been fitted were not functioning effectively.

NATURAL VENTILATION

Since bathrooms are usually private spaces, many only have small windows. But it doesn't need to be that way, and with clever angling or screening, bathroom windows can be designed to be large enough to be opened for good natural ventilation while letting in views of the garden or sky. Bathing or showering can be a pleasant, relaxing experience, and with the addition of a natural view, it can develop a meditative aspect that helps make it more of a daily ritual than a daily function.

Ideally, a bathroom window should be open at least a small amount as often as possible. If you can fit security stays to the window/s and leave them open year-round, it will lower the moisture level in your bathroom considerably.

MECHANICAL VENTILATION

Mechanical bathroom ventilation can be part of a larger balanced-pressure heat-recovery household system, or can be a standalone extractor fan that vents to the outside of the house. Some bathroom ventilation systems come with two extraction ports fitted to the same exhaust, so if your bathroom is near a laundry or separate toilet that also needs extraction, both can be fitted at once. Extractor fans with built-in sensors that automatically switch on or off at a predetermined level of humidity are also available.

Humidity is related to temperature — warmer air holds moisture better than colder air does — so if you heat your bathroom, your ventilation will be more effective in removing moisture. Extraction fans also work better when there is a source of fresh air to draw into the room to help the air move through. If you can partially open a window during or after your shower to let in fresh air while the fan is still on, your fan will work at its best. However, if the outside exhaust vent for your bathroom extraction fan is close to the bathroom window, it may be better to leave it closed (so that expelled air isn't being drawn back in through a window) and open the bathroom door a crack instead.

Always close the bathroom door after you've finished your shower and left the room, so that the remaining moist air is ventilated out through the window or extraction fan, and doesn't move into the rest of the house.

TOILETS

Flushing the toilet is the third largest consumer of water inside the home according to Watercare (after showering and laundry). Older-style toilets use around 12 litres per flush. If you have one of these large-tanked beasts, it's time for an upgrade to a newer dual-flush toilet that uses around 6 litres for a full flush. If you can't afford to upgrade yet, fill a 2-litre plastic bottle with sand and put it in your tank after the next time you flush, which will reduce the amount of water that fills the cistern.

Some of the newest low-flow toilets rated by WELS as 4 stars use as little as 4.5 litres for a full flush and just 3 litres for a half-flush. If you're installing these very low-flow systems in a new home, BuildingGreen advises that long-run horizontal drains may be more prone to clogging with these systems. A slight downward slope on the piping will help avoid any issues. Talk to your architect, builder or plumber about whether they've experienced any issues.

RAINWATER AND GREYWATER

Because water for toilet-flushing doesn't come into human contact, you can use rainwater or greywater. Rainwater is collected from roof run-off in a tank and can be plumbed into your toilet, laundry or garden-hose systems. Rainwater systems are widely available throughout the country (and may be the only available type of water system in further-flung places). Greywater is lightly used water from your shower, washing machine or bath that can be collected in a tank, treated with chlorine and connected to your toilet tank (with a backup mains system) for use when flushing. They've been approved for urban use in some areas only recently, so they're not yet common. Greywater can sometimes be used for garden watering, but it's not suitable

The 339 House uses band-sawn Lawson cypress to line the bathroom, with a steel-trowelled-concrete shower lining. The floor-to-ceiling window in the shower area opens to let steam out directly — a louvred shutter can be pulled across to give more privacy if desired. The bench seating to the left is a concealed bath — the lid lifts up to reveal it. Photo: Simon Devitt

for washing machine use or any kind of use that involves direct contact with humans. See more on these systems in Chapter 13: Utility Spaces.

COMPOSTING TOILETS

Waterless or composting toilets are a tough one for most people to get their heads around. Many Kiwis have held their noses and waved away flies at campsite long-drops, and picture composting toilets as slightly more civilised versions of that. The reality is quite different — modern composting toilets are often less smelly than conventional toilets (a small fan extracts all odours) while using very little or no water.

The second thing people struggle to get their heads around is that human waste can be safely converted into compost through a natural system. Most people believe that human waste is rife with pathogens that can only be killed by strong chemicals. That's not true. Composting human waste at high heat levels kills pathogens, and although composted 'humanure' shouldn't be used on food crops, it can safely be returned to general soil.

Unless you're building in an area that operates on a septic system or are building a Tiny House, you'll be unlikely to be able to put a composting toilet into your home, which is a pity, as their technology has come a long way in the last couple of decades. The Bullitt Center, a six-storey office complex in Seattle that was constructed as a Living Building, successfully uses composting toilets throughout the facility.

THE STUFF WE DON'T TALK ABOUT

Toilet paper

Around the world, every day 27,000 trees are felled so they can be flushed down our toilets — that's 10 per cent of all the paper we use. Of course, harvesting and replanting trees is a sustainable process, as long as it is done well. That's why it is important to choose paper that is either recycled or certified as sustainable by the Forestry Stewardship Council (FSC), the Programme for the Endorsement of Forest Certification (PEFC), or New Zealand's Environmental Choice.

Alternatives to paper toilet rolls started coming onto the local market around 2012. Several brands now offer paper made from bamboo or sugarcane fibre at a price point similar to conventional paper. Bamboo is a grass, not a tree, so regenerates far more quickly, although land still needs to be cleared for its cultivation. Sugarcane fibre is a waste product from the sugar industry, making it very eco-friendly.

Menstrual cups: making red days green

Most women can come up with a ready list of complaints about the inconvenience and cost of their periods — and the planet has a few complaints itself. But because of the reluctance to discuss it, the effects of disposable menstrual products are a largely silent crisis.

The average woman uses 12,000 single-use menstrual products in her lifetime. These are resource- and chemical-heavy to produce, package and transport: the average sanitary pad contains the same amount of plastic as four shopping bags. If disposed of into landfill, they take centuries to decay, and if flushed down toilets (warning: put down anything you're eating immediately) can congeal into hazardous 'fatbergs' that clog sewer systems and drains. In 2013, a 15-tonne fatberg of rotting food and sanitary products was removed from a sewer in London. It was the size of a bus.

Reusable silicone 'menstrual cups' and washable super-absorbent 'period pants' can make disposable sanitary products a thing of the past (and save up to $16,000 in lifetime costs). Menstrual cups can take a little getting used to, but after the first few goes are as easy to use as tampons — plus, you know you are never going to be caught short by an empty box again. 'Period-proof' washable undies that can work as a backup to menstrual cups and absorb the same amount of flow as two tampons. Their designs are modern and even chic — certainly a lot chicer than those old undies stashed in the back of the drawer. A menstrual cup will set you back $30–50 and to purchase a full 'set' of period pants that will see you through a week will be $150-200 — an investment, but if you go through 30 tampons and a pack of pads each period, it's one that pays itself back in a year.

At present, regional council regulations throughout New Zealand state that if a property is or can be connected to the existing wastewater systems, it must be, except under special circumstances. Joel and Rochelle Payne, who are building a Living Building Challenge home in Beachlands, East Auckland, have been approved for an exemption for their home in a test case for urban use, which if successful should make it easier to gain consent for on-site wastewater management.

LINING YOUR BATHROOM

As one of the dampest rooms in the house, bathrooms must be carefully lined to prevent structural moisture and degradation, using materials that are designed to function well in wet-area conditions. The Building Code states that surfaces in wet areas must be impervious, easily cleaned, and have ventilation. All fixtures and fixings must be watertight to prevent water seeping into concealed spaces and causing damage. In addition, if you have a timber floor in the bathroom, it must have an impervious lining to prevent pooled water causing damage.

If you're substantially renovating a wet area like a bathroom, the easiest and most cost-effective way to do it is usually to remove the existing linings and replace them. If you're relining an older home's bathroom, opening up the walls can bring a few surprises, including mould that needs fixing and pipes that need replacing. Leave some money aside in your budget for this.

Standard plasterboard doesn't stand up well in wet areas, even when painted with water-resistant paint. A wet-area-specific lining like GIB Aqualine will be more durable. It can be painted with a gloss or semi-gloss paint, plastered with a cement-based plaster or used as a substrate for tiles, terrazzo or a secondary lining like wood, cork tiles or linoleum.

BRANZ's sustainable building website (www.level.org.nz) has a good reference guide to wet-area construction requirements and materials.

TILES

Ceramic tiles are made from naturally abundant materials, but can be energy-intensive to fire. They're long-lasting and easy to clean, inert (emitting no VOCs, although the grouting and adhesives used to secure them can emit low levels), come in a nearly endless variety of colours, shapes and sizes, and can be very affordable. Preserving or reusing heritage tiles is also an option if they are removed carefully.

Porcelain, the stronger and heavier cousin of ceramic tile, offers similar benefits but its additional weight makes it more suitable for flooring than wall coverage. Wood-look porcelain is a product with growing popularity — good ones look very similar to real wood, although they're colder to walk on.

Right: When renovating a home in Auckland, Tim Dorrington was asked to preserve existing heritage tiles in the bathroom to retain the eco-friendly character. Photo: Emma-Jane Hetherington

Below: Gina Stevens' Tiny House features a composting toilet. Photo: Jemma Wells

Some manufacturers try to claim that wood-look tiles are more eco-friendly than wood because you don't have to cut down trees, but this is a classic example of greenwashing. Wood is a fully sustainable resource, whereas the embodied energy of porcelain is higher than that of wood. Choose them because they're more durable and water-resistant than wood, not because they're inherently more eco-friendly. When installed correctly, ceramic and porcelain tiles are nearly impervious to moisture, although they can crack if heavy items are dropped on them.

Glass tiles are also made of abundant natural materials, and take less energy to produce than ceramic tiles. They also often contain high amounts of post-consumer recycled content, making them an even more eco-friendly choice. Glass mosaics have a luxurious quality to them, but they are difficult to install, and are not a DIY job unless you're extremely experienced. Like ceramic and porcelain, cracking is a risk. The smooth surface of glass can also make them slippery as a floor tile.

Terrazzo tiles are made from a cement binder mixed with chips of stone, such as granite or marble, or recycled glass. It has a variegated effect. Cement manufacture releases high levels of greenhouse gases, and the product has a high embodied energy. Terrazzo can also be installed as a solid surface poured on-site, which removes the issue of grouting (which can attract mould and needs more vigorous cleaning than the tiles) but it's an expensive choice.

Stone is a hard natural material that does not need firing, but the mining process is energy-heavy. It tends to feel colder underfoot than other types of tiles, so may need underfloor heating to be comfortable in winter. Some stones absorb moisture and are not suitable for wet areas — avoid non-vitreous or semi-vitreous stones, and choose vitreous or impervious ones. Where possible, choose locally quarried stone over imported stone tiles.

If you are planning to replace an existing acrylic or stainless-steel-lined shower with continuous tiling, you may need a building consent to do so. Make sure to check the details of your project with both your builder and your local council.

TADELAKT

Tadelakt is a traditional natural Moroccan lime plaster that is often used for wet areas, such as in Turkish hammams, or bathing houses. The plaster is rammed, polished and treated with soap to make it completely water-resistant — so much so that it can even be used for basins and baths as a completely continuous surface, giving a seamless effect and removing the need for grouting. The plaster can be rolled around corners to create curves, and can be sculpted. Harsh chemicals can deteriorate the surface, so only gentle cleaning products should be used. The surface needs to be rewaxed on an annual basis to maintain its integrity.

Tadelakt is a labour-intensive and highly specialised product to apply, especially in wet areas, and there are few people qualified to do so in New Zealand. Resene makes a tadelakt product called Rockcote, and recommends that any tadelakt installation be done in consultation with their technical department to ensure a good result.

FIBRE-CEMENT PANELLING

Polyurethaned fibre-cement panels are often chosen for bathroom linings. They're durable, mould-resistant, impervious to water and some feature a tile pattern that mimics the look of ceramic tiles without the need for cleaning. Polyurethane releases VOCs, so a bathroom area lined with this product would need to be very well ventilated. Ideally, you would let the room air out for a couple of weeks before first use.

Fibre-cement board takes energy to manufacture, and releases respirable crystalline silica particles when cut. These particles have been associated with an increased incidence of silicosis (lung scarring) and lung cancer in exposed workers, so serious safety precautions need to be taken on-site. Cutting and grinding concrete, ceramic and porcelain tiles also releases silica particles, as does mixing mortar and grout. Wetting these products down during cutting reduces the risk of airborne particles.

WOOD

Wood is a feasible and beautiful option for lining a bathroom (except for the shower areas), giving it a sauna-like feeling, but the wood needs to be sealed with a waterproof coating and the bathroom well ventilated. Solid, dense timbers will be more resistant to moisture absorption and warping than softer woods. Particle board and MDF are not a good choice as they absorb moisture too readily, leading to potential warping and damage, but well-sealed plywood can work as a finish outside of shower areas.

CORK TILES

A highly ecological material, cork tiles can be used in the bathroom as long as they are well sealed. They have the warmth and comfort of timber at an accessible price.

LINOLEUM

Linoleum and vinyl are words often used interchangeably when talking about hard-wearing sheet materials, but the two products are quite different. Real linoleum is made from natural ingredients, including linseed oil, ground cork or sawdust, pine resins and calcium carbonate on canvas or burlap backing. It does not emit VOCs and has a low embodied energy. Vinyl is a manufactured product made from petroleum products and uses large amounts of chlorine in the production process. It does release VOCs, and is energy-intensive to manufacture. If you're talking to someone about linoleum flooring, be sure to clarify whether it is real linoleum, and not a vinyl substitute. Although both can be used as bathroom flooring, linoleum is less water-resistant, and will need periodic resealing to maintain its integrity. Serious flooding could lift it.

CHAPTER 11

THE BEDROOMS

Good sleep has become something of a cultural obsession over the last few years, with everyone from international media magnate Arianna Huffington to actress/lifestyle-website founder Gwyneth Paltrow passionately advocating the importance of eight hours every night. But good sleep isn't just a fad — the science behind the restorative effects of good sleep, including solid chunks of the deep, REM-phase sleep when our brains process memories, is well established. On the flip side, the effects of poor sleep include reduced ability to concentrate at work or school, increased susceptibility to illness, driver reaction times on par with being legally drunk, and even weight gain.

Developing a relaxing bedtime routine that could involve gentle exercises like yoga, drinking herbal tea instead of alcohol or caffeine, and using a meditation app to 'wind down' your brain will certainly help you nod off, but making fundamental changes to your bedroom environment will have an even more profound effect on how well rested you feel when you wake up.

Creating a good sleep environment is known as 'sleep hygiene', which may sound rather medical, but is actually a great way to describe the process of cleaning up the two main kinds of pollution that affect our sleep — light pollution and air-quality pollution.

WINDOW TREATMENTS

Our bodies are governed by circadian rhythms that respond to changes in natural light throughout the day, increasing and decreasing our energy, heart rate, and hormone and brain-activity levels as we move from day to night and back again. The more in-tune your body is with natural light changes, the more reliable your circadian rhythms will be. In practical terms, this means that the darker your room is through the night, the better your sleep will be. It's also important to keep your body in sync with the natural colour changes of the light through the day. As bedroom lights are normally used in the evening, look for lightbulbs with a warm/yellow colour temperature (3000 K or lower). Alternatively, energy-efficient LED bulbs with a colour temperature that can be adjusted from cool (for the mornings) to warm (for evenings) by WiFi remote are available from specialist lighting retailers.

In cities, 'light pollution' from streetlights, car headlights, neon signage, security lights, and commercial and residential lights has been linked to significant effects on health. A wide-ranging study from Stanford University found that people in cities with high night-time light levels slept fewer hours on average than people in rural areas, and were unhappier with the quality of their sleep, resulting in more fatigue during the day. Studies have even suggested a link between high night-time light levels and an increased incidence of cancer.

Curtains and blinds in natural linens are double-layered for warmth in this Wairarapa home by Donnell Day Architects. Photo: Elizabeth Goodall

Choosing well-fitted window treatments that are lined to exclude as much light as possible helps minimise light pollution and improve sleep. Of course, window treatments aren't just about keeping a room dark at night — they also increase insulation value, offer shade and privacy during the day, and add texture and colour to a room.

In the mornings, open your curtains and blinds completely as early as possible to allow natural light into the house. Some motorised blinds have sun sensors that open and close the blinds automatically to make the most of natural light patterns.

CURTAINS

The biggest mistake people make with curtains is not making the proportions generous enough. Curtain rails should be placed well above the upper edge of windows and glass doors, ideally with a pelmet along the top to keep heat in. If you don't want to add pelmets (not everybody likes the look), then hanging the curtain rail as close to the ceiling and as tight against the wall as possible will do a similar job, with the added benefit that placing railings high can give the illusion of extra ceiling height in the room. The rails should also extend far enough out at each side to enable the curtains to be pulled right away from the windows during the day, letting the maximum amount of natural light in.

Curtains should go all the way to the floor wherever possible — not an inch or two above, actually touching the floor — to prevent heat loss. Those extra few inches are really worth the investment — and soft linen, hemp, bamboo and lightweight cotton curtains look particularly lovely when they are given enough length to 'pool' on the ground. This will also help reduce condensation on the windows.

Double or even triple layering curtains with a wadding and/or blackout layer improves their insulation value much more than thermal coatings. If you find that the additional layer/s makes the fall of the curtains look lumpy (especially with lighter-weight curtains), you can get double rails that allow you to hang a thermal/blackout curtain layer close to the window, and a lighter or more decorative curtain in front. Look for curtains made of natural, and ideally organic, fabrics, and prints that use non-toxic inks.

SHADES AND BLINDS

Shades and blinds don't usually add much insulation to windows unless they are double-layered fabric blinds or pleated 'honeycomb' blinds, so they're best on windows that already have insulated glazing, or where controlling the amount of heat and light coming in from the sun is more important than the amount going out. Making sure that blinds are well fitted — going right to the edges of the frame — is key to maximising their insulation and shading value.

There are many different styles available — Venetian, Roman, roller blinds, shutters and louvres, Japanese shōji blinds, vertical panel blinds,

PREVENTING AND CLEANING CURTAIN MOULD

If — like most houses in New Zealand — your windows are single-glazed, you'll be familiar with opening the curtains after a cold night to find the windows beaded with condensation. If this moisture soaks into the curtains, mould can quickly take hold. Condensation should be removed from windows as soon as possible — a handheld window vacuum can do this in seconds, or run a towel over the panes to remove moisture before opening the windows to let air circulate and dry out.

Mould grows on dust and dirt, so regularly wiping windows and their surrounds clean will discourage mould growth. A good splash of white vinegar or a few drops of tea tree oil in the water used to dampen your cleaning cloth will act as a natural anti-fungal agent.

Keep an eye out for dark spots on your curtains, and wash them at least once every winter. Natural fabrics can be washed in the washing machine on a delicate cycle, but curtains with thermal backing need gentler handling. Hand-wash them in the bathtub with a cup of white vinegar added to the water, and dry them by lying them out as flat as possible, particularly the thermal backing, which can stick to itself if it is wrinkled or folded when drying. If you have little dark spots of mould on the curtains that haven't disappeared after washing, don't worry — although the spots are unsightly, the mould spores will have been killed, and that's the important thing.

cellular and pleated, to name just a few. Likewise, they come in a vast range of materials, from natural sources like wood, bamboo, hemp and papers, to synthetic polymer foams, aluminium and flexible plastic weaves. Your choice of style and material will largely be guided by how you want your blinds to perform — whether you're looking for insulation value, partial or full sun and light blocking, privacy or a decorative effect — as well as your tolerance for cleaning them (if you've never cleaned a Venetian blind, count yourself lucky) and the style of your home.

Natural materials will normally be recyclable or biodegradable at the end of their lifespan, though old wooden blinds are likely to have been varnished or treated, so aren't suitable for burning. The important questions to ask with synthetic blinds is whether the fabric emits VOCs, how environmentally

SLEEP TIP:
ELIMINATE TINY LIGHT SOURCES

Even small sources of light can have a disruptive effect on your sleep. If you have a digital alarm clock, baby monitor or smartphone that lights up with notifications, or a glowing standby light on your bedroom television, cover them up when it's lights-out time.

friendly the production processes are, whether the manufacturer accepts returns at end-of-life for recycling, and whether the company is involved in compensation schemes (like CarboNZero) to offset the effects of the production process.

AIR QUALITY

Air quality is made up of three key elements — temperature, relative humidity and pollutants. An ideal sleep environment has a temperature of around 18–19 degrees, relative humidity between 40–60 per cent (for more on humidity, see the ventilation section of Chapter 13: Utility Spaces), and a low level of pollutants such as VOCs and smoke particles, and allergens like dust, mould, pet hair and pollens.

Poor air quality demonstrably affects how well we sleep, how good we feel and how well we perform. A 2015 study published in the journal *Indoor Air* showed that when indoor air quality was improved through ventilation, subjects slept better, felt better the next day, and performed better in a logical thinking test.

Unfortunately, bedrooms are breeding grounds for VOCs, dust and mould, and are often kept too warm at night. Poorly ventilated bedrooms with single-glazed windows, or with an ensuite that has a shower, can experience serious moisture build-up, resulting in condensation on the windows. But there are simple ways to improve air quality when you're renovating or building, or even by making small habit changes.

SIX WAYS TO IMPROVE AIR QUALITY

1. Change your carpet for hard flooring

Carpet builds up dust, dust mites, dander (dead skin from pets), hair, pollens and dirt to a much greater level than hard floor surfaces. The New Zealand Asthma Foundation recommends hard floors as the healthiest option for people suffering from respiratory illnesses — and in New Zealand, that's one in six of us. Although the idea of stepping out of bed onto a hard, cold surface in the morning might not appeal, a luxuriously plush rug, like a sheepskin or a deep-pile wool carpet rug can be a more pleasant surface to step out on to (and easier to shake the dust out of than a carpet) — and just remember to keep your slippers or Uggs right next to the bed. And whether you have carpet or hard flooring, look for a vacuum cleaner with a HEPA (High Efficiency Particulate Air) filter, which removes particles as small as 0.3 of a micrometre in size.

2. Open your windows

Opening a window — ideally as wide as you can — is a simple, free way to ventilate a room. Windows should be left open for at least 15 minutes at a time, although the greatest benefits are from windows left open continuously. If you fit security stays to your windows, they can be left open all day while you're out of the house, or 24 hours a day in the warmer months.

3. Put in a ceiling fan

Fans don't actually lower the air temperature in a room, but the movement of the air across your skin makes you feel cooler by increasing the evaporative effect of sweating.

If you don't use air conditioning, on a hot night it's best to have the fan on and the windows open — closing the windows doesn't 'keep the cool air in', it just means you're breathing recirculated stale air. In fact, since the fan itself generates a little heat (because it's driven by electricity), you may even increase the temperature in the room marginally by having the windows closed.

In winter, reversing the direction of the ceiling fan can help make the room more comfortable by evening out the temperature. If you use an upright heater that is slow to build up heat in the room, turning on the fan will help the room warm up faster. Check the star ratings on ceiling fans for the best energy efficiency.

4. Invest in an air purifier

If you suffer from a serious or ongoing respiratory condition, and don't have good ventilation in your bedroom, an air purifier can help. These devices draw in polluted air, including pollens, bacteria, mould and dander, and clean it through a filter. Some, such as the Dyson Pure Hot+Cold Link, claim to be able to remove nearly 100 per cent of fine particles in the air using HEPA filters. Some air purifiers can be linked to an app to control them remotely, and can also function as a small heater or air conditioner. They're not cheap, starting at $120 upwards, and the Dyson comes in at nearly $1000, although it's inexpensive to run. But if you have chronic allergies, it may be worth the investment for a room you spend a third of your time in.

5. Ditch the scented candles and incense

Before you light that lavender-scented candle next to your bed, consider what it's sending into the air. That dancing orange flame actually means soot particles are being released (cleaner flames burn blue), and the relaxing scent may be artificial, and contain VOCs like formaldehyde, acetaldehyde and acrolein.

If you can't give up your scented candle habit, choose soy or beeswax candles instead of paraffin wax, avoid multi-wick candles, and look for essential oil fragrances rather than artificial ones. Only burn a candle for an hour at a time, and trim the wick before lighting it.

BEDS AND BEDDING

We spend up to a third of our lives — around 25 years on average — asleep, yet most of us invest more in our cars than we do in our beds. And given that we spend all of those hours flat out in direct surface contact with our mattress, pillow, sheets and duvets, sometimes naked, breathing deeply, it's important that the materials you're lying on (and under) aren't doing any harm to your health, or the environment.

MATTRESSES

The mattress market has undergone a transformation in recent years, with the arrival of deliver-to-door online boxed mattress companies offering compression-packed foam mattresses. These rapidly became popular due to their lower prices and generous, multi-month returns policies (allowing people to avoid the universally weird experience of in-store bed testing).

As more companies have entered the market, 'mattress review blogging' has become an industry of its own, meaning that whichever mattress you're considering buying, it's likely that there will be no shortage of reviews and ratings online for it.

Most people buying a new mattress have two criteria in mind: How does it feel to lie on? and How much does it cost? But the first consideration should be: What is this made of? Conventional box-spring and memory-foam mattresses are often made from synthetic, petroleum-derived foams, and can off-gas irritating and allergenic VOCs. Then, they're covered in an outer layer of cotton that is often produced using chemical pesticides and bleaches.

Futon mattresses are usually made from all-natural materials. A raised, slatted platform allows air to circulate underneath. Photo: Simon Devitt

Foam mattresses also contain chemical flame retardants, some of which have been associated with nervous system disruption, and liver and thyroid toxicity. The most concerning of these, polybrominated diphenyl ethers, or PBDEs, were phased out by most mattress manufacturers in the mid-2000s after being banned in the European Union and Canada, among other places. However, older mattresses may still contain them. If you've had your mattress for 10 years or more, it probably needs replacing for support reasons anyway, but it could also contain flame retardants that have now been banned.

There are four types of mattress, each of which has its own environmental and health considerations.

Innerspring

Innerspring mattresses, the most common type in New Zealand, use systems of coiled steel springs to give support, with a padded layer of cotton or foam on top. They're lighter-weight than most memory foam and latex mattresses (thus easier to move around and flip if needed), and generally contain fewer flame retardants, especially if the padding layer is cotton or wool, not synthetic foam. However, the glues used to hold together layers in innerspring mattresses may release VOCs in addition to the VOCs that come from synthetic foams (see page 197).

One 'fact' that floats around internet and newspaper articles from time to time is that innerspring mattresses double their weight over 10 years due to a build-up of dead skin, dust mites and their droppings. Fortunately, this is an urban myth, although it's true that a well-used mattress contains anywhere from hundreds of thousands to millions of dust mites. You can decrease the dust mites and other build-up in your mattress by regularly vacuuming it (after removing the sheets).

Most of the materials in innerspring mattresses can be recycled, which is important, because if disposed of in landfill, they take up large amounts of room and can wrap themselves around moving machinery.

WHAT DO YOU DO WITH A USED MATTRESS?

Mattresses need to be replaced on average every 10 years due to compression and build-up of dust mites and their droppings, dead skin, sweat and other stuff that it's really best not to spend too long lying awake thinking about. Trouble is, at present, most mattresses end up in landfill. Over 300,000 mattresses and bases are sent to landfill in New Zealand every year, according to Auckland Council, which has partnered with a pilot recycling programme called Rebound to try to reduce this wastage. They estimate the programme could eventually recover over 2 million kilograms of metal each year, and over 3 million kilograms of foam, timber and coir. To find out whether your old mattress is suitable, and how to recycle it, go to www.rebound.org.nz.

Foam

Most foam mattresses are made of non-biodegradable synthetic polyurethane foam, which is produced using petroleum. These mattresses can have a noticeable chemical smell, which indicates the presence of VOCs. Some also contain gel microbeads, designed to address the fact that dense synthetic 'memory foam' mattresses don't dissipate heat as well as natural foam or latex mattresses do, making sleep hotter (and encouraging moisture build-up from sweat).

Foam mattresses can be recycled, particularly into carpet underlay. Because synthetic foams are more flammable than natural foams, synthetic mattresses are likely to contain the highest levels of flame-retardant chemicals. If you're considering a synthetic foam mattress, look for international certifications, particularly Greenguard Gold or CertiPUR, which certify lower or zero levels of certain flame-retardant chemicals and VOCs.

Plant-based foam mattresses use foams derived from plant oils (usually soy, sunflower or castor-bean). These mattresses are often marketed as 'natural' or 'eco' mattresses. However, they are almost certainly not made entirely from plant foams — most are a blend of plant and synthetic foam, and may have significantly more synthetic than plant foam in the blend. Plant-based foams do emit fewer VOCs than fully synthetic foam mattresses.

Synthetic foam and plant/synthetic foam mattresses often come with manufacturer's advice to let your mattress air out for a few days (to allow some VOCs to disperse) after unwrapping, so look to take delivery of your mattress a week or more before you need to use it. It's best to air it out in a spare room, or ideally outside if the weather is good.

Natural latex mattresses are made from foam derived from a milky fluid tapped from rubber trees. Rubber trees are grown in plantations that can be certified sustainable, or harvested from natural forest (also known as 'jungle rubber'). Latex mattresses may contain other materials at the core, such as coir (coconut fibre). They're very durable, and often last longer (up to 20 years) than other types of mattresses.

Some mattresses labelled 'latex' may actually be synthetic latex made from petrochemicals, or a mix, so be sure to clarify with the retailer what material the mattress contains. Products that are 100 per cent natural latex can be organically certified, so look for Oeko-Tex, USDA Organic, Global Organic Latex Standard, Cradle to Cradle and Eco-INSTITUT labels, as well as the previously mentioned certifications for synthetic latex or latex blends.

A certified 100 per cent latex mattress with certified organic wool, hemp, bamboo or cotton outer is one of the most ideal mattresses you can buy from a health and eco-friendliness viewpoint. The main drawbacks of natural latex mattresses are their higher upfront cost, and heavier weight, although they don't need flipping like some synthetic foam and innerspring mattresses do.

Futons

A traditional form of bedding in Japan and Korea, in Western countries the word 'futon' usually refers to a flexible padded cotton mattress about half the thickness of conventional foam or innerspring mattresses. These mattresses may be reinforced with wool, latex, coir or polyester padding in the middle.

Futons in Japan are even slimmer and more flexible than the ones New Zealanders are familiar with. They're laid onto tatami mats or a low frame for sleeping (not placed directly on the floor, as this doesn't allow them to breathe properly), and then are aired out and rolled or folded away into a cupboard during the day to free up floor space. In Japan, the word refers to both the mattress itself and the covers and pillows that go with it.

Organic cotton futons, either with or without a padded centre made of other natural fabrics, are an excellent eco-choice. They're highly breathable, wick moisture away easily, and have a level of natural fire-resistance. Futons are firm — which is good in terms of support — but if they're firmer than you're comfortable with, they can be layered quite easily. Given that they're usually lower-priced than standard mattresses, this is a feasible option. Futons are biodegradable, and can also be recycled.

HEALTHY SLEEP TIP: STOP MAKING YOUR BED

Although these words will horrify your mother (and followers of productivity routines who insist on starting every morning with a perfectly made bed), you could be doing yourself more harm than good by making your bed as soon as you get up.

If you've ever straightened your covers after breakfast and been surprised by how deliciously warm your bed still feels inside, the bad news is that that humid warmth is a haven for hundreds of thousands of dead-skin-munching, allergy-causing dust mites (and their droppings).

Flinging the covers right back after you get up and airing out the bed exposes mites to light and air, killing them and disrupting the breeding cycle. Plus, you'll be able to spend a few extra minutes under the covers with the time you'll save on bed-making.

Waterbeds

Waterbeds had a real moment in the 1980s and early 1990s, and they are still around, albeit less popular these days. They're typically made from vinyl, which can off-gas, with newer ones having reinforcing synthetic foam sidings and uppers. They need to be heated year-round, and have a shorter expected lifespan (5–10 years) than most mattresses, as well as being more complicated to move. Some people swear by their comfort and low dust mite levels, but from a strictly eco-perspective, they're a less desirable choice.

PILLOWS AND BED LINEN

Even more so than mattresses, pillows are often chosen on texture rather than material. But wouldn't you rather spend eight hours breathing right next to a natural, non-toxic material than a synthetic foam that may be off-gassing directly into your face? Plus, if your existing mattress is synthetic, a pillow made of natural material can form a barrier to mitigate against possible off-gassing.

Natural pillow fillings include wool (look for products from animal welfare-certified producers), to more vegan-friendly organic cotton, bamboo, kapok, hemp, buckwheat husks (very firm, and popular in Asia), or latex. Some pillows are infused with herbs or essential oils that are thought to promote sleep.

For a while, thread count was the ultimate marker of bed linen quality, but in reality few people can tell the difference between 200 and 800 threads per inch. The choice of material is likely to be more important to a good night's sleep in the long run. Organic cotton, hemp, bamboo, silk and linen (which is made from flax) sheets, duvet covers and pillow slips, and wool, bamboo or Tencel (the common brand name for lyocell, a fabric made from cellulose fibre derived from eucalyptus wood pulp) duvet inners are naturally breathable. Look for products that have an organic production certification, such as the Global Organic Textile Standard (GOTS) or Oeko-Tex.

Mercerisation is a fabric treatment that is often used on bed sheets. It makes material stronger, softer, easier to dye and more pearlescent. It's mainly used for cotton, but hemp and linen sheets can also be mercerised. Mercerising involves soaking the material in a sodium hydroxide bath, then neutralising it with an acid bath. Residual chemicals can remain on the fibres, so always wash new bed sheets well before using them.

BED BASES

The two main types of bed base are platform and slat bases, both usually made of wood or metal. Platform bases have a solid base on a frame that may be padded or covered with fabric. The base may be able to split into two to make it easier to move, or be fixed on castors. Slat beds have fixed or moveable slats set across an open frame.

Platform bases may be held together or have the padding attached using

This airy walk-through wardrobe allows air to flow freely around the clothes. Wood has moisture-absorbing qualities that helps combat mould growth. Photo: Simon Devitt

glues that can off-gas VOCs. Look for bases secured with nails or screws rather than glues. Solid bases also don't allow air to circulate as freely as slat bases do. If you have a solid base, flipping your mattress regularly becomes more important to avoid build-up of dust mites and moisture.

The ideal bed base is made of solid untreated timber (rather than particle board), sourced from a certified sustainable plantation in New Zealand, finished using a plant-based stain, paint or oil. Metal bed bases should be used for as long as possible and sent to a charity store or scrap metal dealer for recycling when you have finished with them.

WARDROBES

The clothing world has come under intense scrutiny in the last decade for the wasteful and toxic processes and worker abuses associated with the rise of the US$110 billion fast-fashion industry. Fast fashion — cheap, on-trend garments, often with a high polyester content, produced in large-scale factories in developing or Third World countries — is creating a resource-consumption and waste crisis. The textile industry uses huge amounts of water, energy and chemicals for fertilisers, pesticides and manufacturing, and generates hundreds of billions of kilograms of greenhouse gases and water and air pollution — and it's only getting more dire.

A 2016 Greenpeace fact sheet says global clothing production doubled between 2000 and 2014, while at the same time, the average amount of time a person keeps an item of clothing halved. In the past, people would repair or repurpose items of clothing that showed signs of wear and tear; now, they're simply replaced by something more 'current'. *The Guardian* reports that the average lifespan of a garment in the United Kingdom is now 2.2–3.3 years from purchase to discard.

Although some big retailers are encouraging customers to bring back used garments for recycling, in reality it's difficult to recycle used clothing, particularly cheap, poor-quality garments made from synthetic fabrics or poly-cotton mixes. Clothing in a reasonable state can be sold in second-hand or vintage stores, or sent to developing countries, extending its lifespan. In 2014, 4.3 million tonnes of recycled clothing was traded around the globe, which sounds pretty good until you compare it to the 14 million tonnes of clothing that Americans alone throw out every year. However, textile recycling, which is the real key to managing the effects of the industry, is more difficult. Natural fibres like wool and cotton downgrade in the recycling process, and are often not good enough quality to be made into new garments. Although companies are experimenting with processes that can separate out synthetic and natural fibres, nothing is yet commercially viable.

So, the most important environmental aspect of your wardrobe is not so much the structure itself — whether standalone or built-in — but keeping the clothes contained within it to a minimum. A helpful side-effect of reducing the number of clothes in your wardrobe is that it allows air to circulate more freely within a space that is typically not well ventilated, reducing the common problem of moisture build-up and mould growth. Always make sure your clothes and shoes are absolutely dry before putting them away.

If you're designing a new home, consider incorporating a walk-in or walk-through wardrobe that will allow air to circulate well. A small window that can be left open will also improve ventilation and supply natural light. For traditional built-in or freestanding wardrobes, leave the doors open often. If your wardrobe tends to be musty, you can put a commercial moisture absorber in the back at floor level. Despite the typically garish packaging, these are based on a natural substance — calcium chloride crystals — and are non-toxic.

A part of the decluttering movement, the 'capsule wardrobe' concept is a modern interpretation of the old-fashioned idea of building a classic wardrobe of high-quality staple garments that will last a decade or more. A typical capsule wardrobe contains 30–40 seasonal pieces, including shoes and accessories, but excluding sleepwear, underwear and workout clothing. Look online for 'Project 333' wardrobes (33 items rotated over 3 months) and you'll find plenty of 'minimal style' inspiration.

CHILDREN'S ROOMS

Built-in storage and flexible-use spaces paired with natural materials make this Titirangi kids' room by Dorrington Atcheson Architects eco-friendly. Photo: Emma-Jane Hetherington

Creating a sleep-healthy environment in children's bedrooms is crucial, especially given that their lungs are still developing. New Zealand has one of the highest rates of childhood asthma in the world, which can interfere with sleep, cause missed school days, and require treatment with corticosteroids.

Children are also clutter-generating machines, and their constant growth and changing needs mean they leave a trail of outgrown clothes and toys in their wake. As much as possible, choose furniture and toys made of sturdy natural materials that will last for years and be able to be passed on, rather than thrown away, when your kids are done with them.

FIVE RULES FOR ECO KIDS' ROOMS

1. Use low-VOC paints

A fresh coat of paint is one of the best ways to update a bedroom. But always use low- or zero-VOC paints to avoid irritating young lungs. If possible, paint the room at least a week before your child will sleep in it, leaving it well ventilated to give any VOCs time to disperse.

2. Choose a latex or futon mattress

Avoid the issue of off-gassing by choosing a cot or children's mattress made of natural materials.

3. Buy to last

Toddler beds and tiny chairs and tables are undeniably adorable, but within two to three years you'll need to upsize. Look instead for pieces that will accommodate your child's growth, or that can be repurposed once your baby is bigger, like a changing table that converts to a clothes-drying rack, or a coffee table that can be used as a kids' table while they are little.

4. Keep the clutter in check

Cluttered surfaces equal dust, and good storage is the key to keeping surfaces clear. Under-bed or under-bench seating drawers, in-wardrobe storage systems, wall-mounted shelving, soft toy nets and cubed shelf systems with pull-out storage boxes will give you an easy place for everything. Remember: the harder something is to put away, the less likely a kid is to do it, so give everything a clear home (and label it). Vintage trunks and wooden crates can make great 'shabby chic' recycled storage solutions.

5. Bring nature in

Studies have shown that childhood is a crucial time for developing a love for the natural world and a passion for caring for it. A plant, fresh flowers, or interesting leaves, rocks, feathers or shells put on display keep kids in touch with the outside world.

A variety of houseplants give a lush feel to this Auckland 'jungalow' apartment — but may do more to combat noise pollution than air pollution. Photo: Ron Goh

HOUSEPLANTS ARE GREAT FOR PURIFYING AIR . . . AREN'T THEY?

You'll often read articles recommending peace lilies, spider plants, English ivy, aloe vera and weeping fig as powerful natural air purifiers that absorb VOCs through their leaves and microbes in their soil. But although it's true that indoor plants do absorb benzene, formaldehyde and other chemicals that off-gas from common household products, you would need a jungle in your bedroom for it to make any significant difference to your air quality. On the other hand, plants do add moisture to the air by releasing it through their leaves, so if you have several large pot plants in your bedroom, every time you water them, you're adding almost the same amount of moisture to the room.

This isn't to say that you should evict your butterfly palm and maidenhair fern immediately. Plants absorb carbon dioxide, generate oxygen, help absorb or refract noise and give a sense of well-being and connection to nature. Just don't expect them to do the work of a HEPA filter.

CHAPTER 12

THE HOME OFFICE

Left: At the Lynds House, the space underneath the stairs was designed to house a casual work or study nook. Photo: Makers of Architecture

Below: Strachan Group Architects' 339 House uses an internal window to provide natural cross-ventilation between the home workspace and the double-height living space. Photo: Simon Devitt

Over the last 20 years, digital technology has (depending on your viewpoint) either fundamentally disrupted the workplace or enabled a revolution within it — or more accurately, outside it. Freelancing and remote working (also called telecommuting or teleworking) have skyrocketed in the last two decades. A 2016 study by Edelman Research showed that 35 per cent of the United States workforce works in the 'gig economy' as freelancers, temps or contractors, with the majority of them working that way by choice, not necessity. And in New Zealand, an MYOB survey revealed that 46 per cent of small or medium-sized businesses have workers who spend part or all of their week working from somewhere outside the office.

A similar result was found in the 2013 Regus Global Indicator, which surveyed 26,000 business managers around the world and found that 48 per cent of them worked half or more of the week remotely. Millennials are more likely to prefer freelance or remote work, but there's also a growing contingent of Gen X parents juggling work and childcare, mid-life entrepreneurs and senior creatives who appreciate the flexibility of working from home at least part of the time.

Aside from the benefit of flexibility for the employee or business owner, working from home can have considerable environmental advantages — most obviously, the removal of commuting time, and the related reduction in transport-based pollution and fossil-fuel use. Teleworking has also been reported to boost productivity by more than a quarter, improve worker motivation, reduce distractions and lower costs for employers (like the traditional desk for every office worker).

The Australian government likes the idea so much that it has set a specific target: to have 12 per cent of the public sector teleworking by 2020. In New Zealand, employers are bound by the Flexible Working Arrangements section of the Employment Act to give consideration to worker requests for flexible working, which can involve working part or all of their hours from home. That means more of us are likely to find ourselves with the option of working from home — and having to find a space within our home to do it.

With the upside of teleworking and freelancing comes a downside. Some of the benefits that accrue to employers from offering flexible working can end up being borne by the employee. In working from home, workers become responsible for the ergonomics of their own workspace, and may find themselves expected to carry the cost of 'incidentals' like their own power, water, connectivity and rental that would otherwise be carried by the company.

Regularly working in a space that is not well set up can even cause as many problems as it solves. Your home office needs to be lighted well, heated in winter, cooled in summer and well ventilated year-round, all of which use energy. If these are done less efficiently than they would be by the company, the net cost to the environment might be greater than if the worker simply caught a train to a well-set up office. Poor ergonomics — uncomfortable chairs, computer setups that put stress on necks and wrists, lighting that causes eye-strain — can have negative health effects and diminish the ability to focus. This is particularly important for home offices that are also used as study and homework rooms for growing children and teens.

As someone who has worked from home for more than a decade, I'm in the camp who could never go back to working in an office full-time. But it takes time and a series of adjustments to set up a home office space that's efficient, environmentally friendly, and comfortable for the way that you like to work.

THE RIGHT LOCATION

If your work is intermittent or very part-time, you may be able to fit a desk, chair and a little shelving into unused space within the house, such as a nook under the stairs or in the hallway, or an area within your living room. But if you're planning on working from home for half a week or more — particularly if you have kids who regularly want your attention — you'll do much better in a separate room that is quiet and able to be closed off for times you need to focus or make important client calls. If you don't have room for a dedicated space, a guest room or even your own bedroom can do double-duty as a home office.

Look for a space that has a window that looks out to a green view, or one where a green view can be created — for example, a fence or wall that could have ficus or ivy grown up it, or bamboo screening grown in front. A study by the University of Illinois Department of Landscape Architecture found that students in a classroom with a green view (rather than no windows or a view of a built environment) performed better on tests requiring focus, and recovered more quickly from stress. If there's no option for a green view, add plants within easy view (able to be seen without turning around).

LIGHTING

The ideal workspace has enough natural light to minimise or eliminate the need for artificial lighting during the day and to give the worker a sense of connection to the outside world. Daylight also helps to regulate our circadian rhythms, improving sleep and increasing a general sense of well-being. Dim rooms make our eyes work harder and can reduce energy levels. However, it's also possible to have too much of a good thing with daylight — too much bright light in a room creates contrast and glare that are hard on the eyes, and can overheat the space (particularly in a smaller room like an office), leading to fogginess and difficulty concentrating.

Daylighting in a big, multi-person workspace like a company office is very difficult to get right, because different people have different levels of light sensitivity. A room that's bright and cheerful to one person may have too much glare for another. Fortunately, ideal light levels can be much more finely tuned in a home workspace.

High windows like clerestories or skylights away from the work area that add general brightness to a room without beams that fall directly on the desk or computer monitor can be ideal. Computer screens should be set up perpendicular (at a right angle) to a natural light source from a window. Avoid putting computers with their back to a window, as it increases contrast and glare. If that's the only option, a sunscreen blind during the brightest parts of the day will make it more comfortable. Tinted glass can also cut down glare. If you have more than one window in the room, and one is glary and the other isn't, it's okay to put tinted glass in the one that gets the most light through it, and plain glass in the other. Nature isn't symmetrical, so our building design doesn't need to be either.

The paint colour you choose for your walls can also have an effect on the comfort of the space. If there's a big contrast between the colour of your wall behind your computer and the brightness of the screen, your eyes have to work harder. If the room is low in natural light, white or pale-coloured walls can reflect what light there is, increasing the perception of brightness.

TASK LIGHTING

Where daylight is insufficient, a well-directed desktop task light works better (and is more energy-efficient) in an office space than switching on an overhead light.

Desktop task lights that are articulated at the base, 'elbow' and head allow you to focus light exactly where you need it when reading printed documents or working on paper, and direct light away for a more diffused effect when working on a computer.

There are beautiful, New Zealand-made articulated desk lamps from local brands Resident, and Douglas and Bec, as well as dozens of international designs. LED bulbs will mean that the desk lamp doesn't become a distracting source of heat.

FURNITURE

Furniture is one of the biggest costs of an office (unless you have expensive tastes in technology). Second-hand office suppliers can offer great prices on recycled office furniture in good condition, but your options in terms of brands will be limited, and it may be harder to trace the materials and authenticity of items sold as easily as it is for well-known designer pieces. If you can afford to invest in well-designed new items, the payback on doing so is pretty easy to justify, especially if you work from home full-time. Plus, it's a business expense.

DESKS

Spending eight hours a day seated at a desk has been described — perhaps a little hyperbolically — as 'the new smoking' by Dr James Levine of the Mayo Clinic. Dr Levine is so passionate about combatting the hunched posture, organ compression, muscle degeneration, increased obesity and poor circulation of extended sitting that he invented the treadmill desk, an invention that it's safe to say is unlikely to catch on widely. Still, there's a middle ground that is growing in popularity: the standing desk.

Most standing desks are, in reality, both sitting and standing desks — they're adjustable. That's because standing all day carries its own set of health issues, from muscle fatigue to varicose veins and numb feet. So choosing a standing desk that's easy to adjust to a sitting version every day is essential. The height-adjustable Humanscale Float desk, available in New Zealand, was recently awarded Living Product Challenge certification by the Living Future Institute, meeting a strict list of eco-friendly criteria from production to materials to recyclability.

For a space with lower light levels, a white desk surface can gently reflect light, creating more balance in the contrast of the space. But in a room with high light levels, white surfaces can cause glare. A wooden desktop may be easier on the eye, and may be a more matte finish than reflective melamine (a common desktop material). Look for wooden desks and trestle desktops that have been FSC- or PEFC-certified as sustainable. Raw-edged slabs of sealed wood on trestle legs can make a simple and beautiful working surface.

CHAIRS

If you don't have a taste for space-age aesthetics, looking for the perfect ergonomic office chair can be a dispiriting experience, but this is an instance where functionality has to be prioritised. The most ergonomically designed chairs have adjustable heights, armrests and backrests, a swivel base and a seat and back usually made of some kind of breathable mesh fabric. The Herman Miller Aeron chair, released in 1994, is the original super-modern office chair, and many hundreds of variations on the design have followed in its wake.

Vitra Tip Ton chairs, made of 100 per cent recyclable polypropylene, tilt forward to encourage the same kind of body position as standard ergonomic chairs. They have a pared-back aesthetic that may appeal to those with more minimalist tastes.

Kneeling chairs are an option for people with lower-back problems, particularly coccyx pain. Wooden versions with padded fabric seats and knee-rests are widely available.

STORAGE AND ORGANISATION

Good systems are the key to having a working space that functions well, is easy to navigate and doesn't become an eyesore (particularly if it is part of another living space). Modular shelving and box systems — such as those made by New Zealand company Lundia, which uses sustainably grown New Zealand timbers — will fit into most spaces, and are easy to assemble and disassemble later.

The 'paperless office' is a modern environmental aspiration, and digital technology is making it considerably easier to come close to (100 per cent perfection may be impossible to achieve for a number of types of work). Send invoices, contracts and receipts by email or through an accounting system, and request your suppliers and service providers do the same. Replace paper forms with fillable PDFs or if you have your own website, online forms. If you must print, print double-sided, or use the reverse side of printouts for scrap paper. Use 100 per cent recycled, chlorine-free and/or sustainably certified printer paper.

Have separate bins for used paper, recycling and other office waste. Important documents that need to be shredded can be added to household composting in limited quantities.

Printer ink cartridges can and should be reused and recycled — a new one uses 3 litres of oil in its manufacture, and can be reused many times before it needs to be recycled. In addition, cartridges disposed of in landfill can leak toxic chemicals. Most printer companies offer a take-back service, but the locally operated company Cartridge Rescue will collect packages of 10 cartridges or more for free.

While you are reducing the amount of paper you use, you could also reduce the number of pens that you throw away. Bic alone ticked off its 100 billionth sale of the disposable ballpoint pen back in 2005, and the American Environmental Protection Agency estimates that in the United States alone 1.6 billion pens are thrown away every year. Investing in a refillable rollerball or fountain pen not only makes writing and doodling a more pleasurable experience, it also saves on waste.

DESKTOP OR LAPTOP?

If you want your computer choice to reduce energy consumption, pick a laptop over a desktop. Laptops use as much as 80 per cent less energy than a desktop computer. Besides, what's the point in working from home if you can't pick up your computer and take it to the nearest café or park for a couple of hours of fresh scenery?

With the explosion in use of personal devices over the last two decades has come an explosion in electronic waste, also known as e-waste. This isn't just due to an acceleration in the capacity of digital technology and more people owning more devices, but the fact that technology brands employ a strategy of planned obsolescence, making devices less robust than they could be (and expensive to repair), and introducing a paced schedule of upgraded models every year to make last year's devices seem hopelessly out of date.

Local communications providers also entice customers with offers to trade in their old phone for a discount on a new one, or to sign a contract to 'get the latest phone every year'. This has resulted in estimates that nearly half a billion discarded electronic devices are hanging around in drawers and cupboards across the United States, unused, but not disposed of or recycled. In America, e-waste is the fastest growing segment of the municipal waste stream.

As phones, tablets and laptops become thinner and more complex, they also become more difficult and expensive to deconstruct for repair and put back together again. At the end-of-life, however, they still have considerable value — e-waste contains materials that are both valuable and environmentally expensive and inefficient to extract, like gold, silver and other precious metals, as well as toxic substances that can leach into landfills, such as cadmium, lead, chromium and mercury.

Working from home usually calls for the purchase of a home computer, a mobile phone and a printer/scanner. Depending on your field of work, additional devices might be required, or at least be tempting. When choosing, purchasing and using electronic products, you can reduce your environmental impact in four key ways.

Buy less: Keep your devices for longer — resist the temptation to upgrade just because a newer version of your device is available. Combine as many functions as possible in one device — don't buy a music player as well as a phone; purchase a combined printer/scanner/copier; ask yourself whether you really need a laptop and a tablet or whether buying a keyboard attachment for your tablet will do the trick.

Buy smart: Technology companies have some of the cleverest and most well funded marketing departments in the world. It can be difficult for the consumer to see past the hype and the greenwashing, and find good information on which companies are truly prioritising sustainable production, ethical supply chains, and energy and chemical reduction, and which ones are just the best at selling bells and whistles. Greenpeace publishes an annual Guide to Greener Electronics that digs deep into the multiple issues around communications technology production and gives companies a report card on energy and resource consumption, and the use of hazardous chemicals in manufacturing and in the products themselves. Search for the latest edition online.

Pass it on: If you're purchasing a new device while the old one is still in working order, scrub the old one clean (not with a brush and soap, but by removing all personal data and reformatting the hard drive) and donate it to a community organisation or school that needs it.

Recycle it: If the item has finally given up the ghost, or is so old as to be difficult to work with current operating systems and software, send it to an e-waste recycler for disposal. There are a number of these in New Zealand — try www.ewaste.org.nz, www.ecomatters.org.nz or www.sustaintrust.org.nz. Most of these organisations charge a small fee to cover the costs of sending the items on to an appropriate recycler overseas. Some schools and charities also run collection drives for electronic equipment, as when these goods are recycled at a large scale, some recyclers offer payment for them.

ENERGY SAVING TIP: USE A MULTI-PLUG

Use a single multi-plug to plug in all of your home office equipment — computer, printer, modems, desk light, fan, heater — and get into the habit of switching it off when you leave your desk at the end of the day, to save all of them drawing standby power.

CHAPTER 13

UTILITY SPACES

A suspended drying rack in the well-ventilated laundry takes care of lightweight items like tea towels and linens. Photo: Elizabeth Goodall

Holes in the concrete slab wall of the 339 House garage allow air to ventilate through the spaces. Photo: Simon Devitt

Like the inside of your walls, garages, workshops and laundries are the unglamorous workhorses of a home. Because they need to be extremely durable — water-resistant, stain-resistant, chemical-resistant — and often have high storage requirements, the materials used for them can tend towards having higher embodied energy or carbon footprints. These spaces are often where technological and energy systems (such as solar-power inverters and batteries, 'smart-home' gateway systems, hot-water cylinders and underfloor heat-pump systems) are kept.

SMART-HOME TECHNOLOGY

'Smart-home technology' is a broad term. In the loosest sense, it can mean a single household item that can be connected to the internet, like a heat pump with an app to turn it on and off remotely. Some smart-home installers even categorise installing USB charge points and data sockets as smart-home technology. At the complex end, it means a networked system of devices and appliances that are designed to work together to provide a range of services to the homeowner. These devices are part of what's called the 'Internet of Things', a network of devices that contain sensors, remote controls, software and other electronics, which connect to the internet and exchange data with other devices. Up to 30 billion of these devices may be in use by 2020. Apple,

Google and Amazon are all moving into the smart-home market with 'smart speakers' that work with their online personal assistants like Siri, Google Assistant and Alexa.

Right now, even so-called smart homes are not smart in the sense that they make decisions by themselves — the actions the devices take usually need to be activated, requested or approved by the homeowner in some way. You might have read about fridges that can 'order food for you' when you run out. There are indeed fridges that do this — the Samsung Family Hub is an example — but at this stage the fridge-owner still needs to personally check what's in the fridge (either by standing in front of it with the door open, or by remotely checking in-fridge cameras on their phone or other digital device) and put together a digital shopping list on the fridge or app to be sent to their local online supermarket. The fridge won't actually notice that you're out of milk and eggs, and send a top-up order to Countdown by itself. But that's likely to change in coming years.

Some of the more common smart-home technologies are remote sound, lighting, entertainment system and temperature control, automated and remote control of access points like gates and doors, and energy monitoring systems. Individual switch timers can also turn on and off appliances like towel rails, driers and washing machines at pre-set times to optimise free solar power usage. Some systems also offer the option of electrical fire protection devices and electric vehicle charge stations for indoors and outdoors.

These systems have the potential to improve the energy efficiency of our homes through automatic action. Motion-sensor lights can switch off automatically when a room is empty, dimmer switches can turn down LEDs (decreasing their power consumption) and multi-plugs on timers can decrease the amount of power pointlessly consumed by 'vampire appliances' — appliances that continue to draw a low amount of power even when not in use, such as cordless phones, DVD players, televisions, laptop computers on charge, games consoles and printers/copiers. Some systems have 'goodbye' functions that allow you to switch off lights, heating, entertainment systems and so on en masse as you leave the house.

Just as important is the ability of these systems to monitor your water and power usage as well as indoor temperature and humidity, and provide feedback that can help you improve your home's energy efficiency and environmental health. These can be built into a whole-house system, or be standalone devices that wirelessly monitor usage or act as intermediary plugs to monitor consumption.

If you're putting in a smart-home device (or repairing or changing a big-ticket item like a hot-water cylinder), consider using an electrician who has been trained in the EcoSmart programme. As well as being able to perform all standard electrical repairs and installation, they can advise you on improving your energy efficiency and how to choose the optimal devices for your home. You can find them through www.ecosmartelectricians.org.nz.

GARAGES AND WORKSHOPS

Electric vehicles (EVs) are still relatively uncommon on New Zealand roads — there were only around 5000 registered in New Zealand at the beginning of 2018, with a few hundred new ones being added each month — but the government has set an ambitious target of having 64,000 registered by 2021. By 2022, the cost of purchasing a new small electric vehicle is expected to be lower than purchasing a similar petrol or diesel vehicle, so the tipping point for uptake may come soon after that. With the range that EVs can travel on a single charge doubling rapidly, the average EV is expected to be able to go 500 kilometres or more at a time by the time the cost becomes comparative to petrol or diesel.

YOUR BIGGEST TRANSPORT CRIME: FLYING

New Zealand's poor urban public transport networks and suburban sprawl can make giving up the private car a difficult proposition even for people living in cities. If you're looking at other ways to decrease your transport carbon footprint, your best bet may be reducing the number of flights you take. New Zealand's isolated position in the world means racking up substantial carbon emissions every time we take to the air.

On average, New Zealanders are responsible for 7.65 metric tonnes of carbon dioxide per capita every year. Carbon emissions calculators vary in their estimates, but one recommended by *The Guardian* puts a return flight from Auckland to Sydney at around 0.4 tonnes of carbon dioxide. That means that for every return flight you take, you're increasing your average annual emissions by just over 5 per cent. A jaunt to Los Angeles will create 2.2 tonnes of emissions, and if you're going for a holiday to visit friends and family in England, at almost 4 tonnes of carbon dioxide, you're increasing your average annual emissions by more than half again.

Some airlines, including Air New Zealand and Qantas, offer the ability to offset the carbon from your flights when you book. It's not cheap, adding around $140 to the cost of a return flight to London, but it is a way to mitigate the damage caused.

Overseas, governments are taking bold strides to encourage EV uptake. In Britain and France, the sale of new petrol and diesel cars will be banned by 2040, while Germany has taken aim for a decade earlier. Paris, Copenhagen, Oxford, Madrid, Athens and Mexico City are introducing city-wide bans by 2025, and China has announced that it is considering a ban on the production of fossil-fuel vehicles. Norway, which like New Zealand runs mainly on hydropower, is the world leader in EVs, with 40 per cent of new cars sold being electric.

Increasingly, fast-charge EV connections (and wiring for solar power) will become part of our new home infrastructure. Jennian Homes, one of the largest group home builders in New Zealand, announced in 2016 that all new homes built by the company would come pre-wired for solar power access and EV charging. If you're planning a new home, consider where you might place indoor and outdoor fast-charge stations for the future EVs in the household.

EV technology will dramatically change our transport landscape over the coming decade, and make a significant contribution to decreasing carbon emissions. With the rise of car-sharing networks, Uber and similar apps, and driverless cars as a reality on the horizon, the way we think about cars and car ownership is also changing. The Kiwi tradition of the two to three car household may turn around, with garage sizes seeing a corresponding downsizing.

LAUNDRY

The average American family does six to eight loads of washing a week, according to Procter & Gamble. That's 312 to 416 loads on average a year. Paying a little attention to how you're doing your washing, and the kind of machine you choose, can save a considerable amount of power and water — and therefore money — every year.

WASHING MACHINES

How much water and energy your washing machine uses is shown on its energy-efficiency and water-efficiency rating (EERS and WELS stickers on the machines), and what type of machine it is. In general, front-loaders are substantially more efficient for both power and water than top-loaders, although there are some top-loaders that are designed to use similar amounts of water as a front-loader.

Most of the energy that washing machines use goes to heating water — a cold wash requires only 10–15 per cent of the energy of a hot one. If you tend to wash on a hot or warm setting, switching to cold washes for the majority of your loads will slash your energy use in the laundry. Modern detergents are designed to clean just as effectively in hot and cold water, even the eco-friendly ones (see more on this later in the chapter). Make sure to look for a washing machine with a genuine cold-water setting — in some machines, the 'cold' setting is actually a minimum of 30 degrees Celsius.

Choosing the right size of machine for your family and your washing habits is important. If you're single or part of a small family, you probably don't need a large 7-kg machine, unless you do all your washing in one or two loads a week. If you tend to do small loads several times a week, a smaller-capacity machine will work better for you, the environment and your budget. If you're a larger family that generates a lot of laundry, choosing the largest size you can fit in your laundry will enable you to do fewer full loads than a smaller machine. It should go without saying that it's better to do a full load than two half-loads. Some machines have load-size sensors or the ability to select your load size so that you don't use a full tub of water for a half-load, but two half-loads still uses double the electricity of a single load.

WHAT IS RELATIVE HUMIDITY?

Aucklanders and Northlanders are familiar with the sticky feeling in summer when the humidity reading creeps up. But what many people don't realise is that indoor humidity can get just as high (or higher) in the winter without the air feeling sticky at all.

The reason is that warm air holds a lot more moisture than cold air does. A cubic metre of air that's at 100 per cent humidity at 0 degrees Celsius contains 3.77 grams of moisture. The same cubic metre of air at 100 per cent humidity at 20 degrees Celsius contains 14.62 grams of water. There's about 25 cubic metres of air in an average-sized bedroom, so drying even a small load of washing in a poorly ventilated room can easily get the relative humidity to 100 per cent in winter without the air feeling damp.

Relative humidity is best measured by a hygrometer. Some dehumidifiers display a digital humidity reading, but this tends to be less accurate. The ideal indoor humidity level is between 40 and 60 per cent, ideally towards the lower end. If relative humidity is too low, the air feels dry and uncomfortable, and can parch mucus membranes, even leading to nose bleeds. When relative humidity reaches high levels, mould can grow rapidly. At 89 per cent relative humidity, mould starts growing within seven days. At 100 per cent, it starts growing within 24 hours. Hygrometers, such as the iButton, can be purchased from home and building supply warehouses, electrical shops and online.

GREYWATER

Greywater is lightly used water that may contain cleaning products, dirt, light grease, hair and even particles of food. It's different to blackwater, or sewerage, which contains human wastes, potential food-borne pathogens or dense oils. Water that goes through your toilet or kitchen sink is classed as blackwater; water that goes through your shower, bathroom sink or laundry is usually classed as greywater (laundry water that has been used to wash soiled nappies would be classed as blackwater). Although you wouldn't want to drink it, once it's been treated, greywater is a perfectly fine, free way to water your garden — in fact, it can even contain beneficial nutrients.

Greywater use is regulated by local councils — some are quite open to the systems being installed, while others don't yet permit their use. All greywater systems should be installed by a professional plumber, and may require a resource or building consent. Some councils also require a maintenance contract to be signed with the company that does the installation of your system to ensure that it's kept in good working order.

DRYERS

The electricity consumption of your tumble dryer will be larger than that of your washing machine, although you probably use it a lot less frequently. There are four types of dryers sold in New Zealand — vented dryers, condensing dryers, heat-pump dryers and combined washer-dryers.

Vented or electrical-resistance dryers — the most common type in New Zealand — draw air into the machine, heat it and blow it through the clothes to evaporate moisture as they rotate in the barrel. The damp, warmed air is then vented outside the dryer. It's very important that this moisture is ducted to the outside of your house and not released inside, or you can end up with mould problems and peeling paint in short order.

Condensing dryers pass warm air through your clothes and remove the moisture through a heat exchanger. The heat is released into the laundry room, and the extracted water is stored in a tank or drained away. They're more expensive than vented dryers — around twice the cost.

Heat-pump dryers have been around in Europe and Asia since the late 1990s but are a more recent arrival here. They contain a small heat pump to heat air to dry clothes. When the warm, moist air is cooled, the water is removed and funnelled away. The same air is then reheated to dry the clothes again. This means no air venting is required. Although they're more energy-efficient than vented or condensing dryers, they're substantially more expensive to buy, so they may not be an economic proposition for your family. *Consumer* magazine offers a good cost-comparison graph for dryers.

Washer-dryers combine a washing machine with a condensing dryer, which releases heat but not moisture into the laundry space. They're most suitable for small laundry spaces, such as in an apartment or Tiny House, and since both functions use water, they can be quite efficient.

LINE DRYING

Outdoor line drying is a great choice for your budget and the environment compared to using a tumble dryer. However, drying clothes on a line inside is not a great idea. A full load of washing can release as much as 5 litres of water into the air as it dries. If you're doing that in an enclosed, unventilated space, it's a recipe for raising humidity and encouraging indoor mould growth. Given that we tend to dry clothes indoors more in winter, when temperatures are lower and relative humidity (see sidebar on page 220) can easily creep up, it's little surprise that New Zealand homes have a serious mould problem. Installing a sheltered outdoor clothes line under the eaves or in a carport is a much better alternative than drying clothes indoors in living areas on a clothes-horse. In a well-ventilated laundry room, drying small, lightweight items like tea towels a on rack can be okay.

CLEANING PRODUCTS

If you're considering a greywater system, now would be a great time to take a second look at the cleaning products you're washing away down your plugholes and drains. Conventional cleaning products can contain salts and harsh chemicals that are unsuitable for greywater because they'll damage plants and animal life within your soils.

Cleaning products designed by major 'eco-cleaning' companies such as Ecostore or Earthwise are designed to be compatible with greywater and septic systems. Look for information on the back of the containers, or on the company's website. Liquid cleaning products tend to be less harsh than powdered cleaning products — although not liquid products that contain bleach (sodium hypochlorite) or caustic soda/lye (sodium hydroxide).

Conventional shampoos, soaps and body gels have more stringent human-contact ingredient standards than cleaning products, so in general don't cause any issues with greywater. (Of course, you may want to choose products made from more naturally derived ingredients for your own health reasons.)

CHAPTER 14

THE LANDSCAPE

The natural landscape of a site is where the idea of a home first truly starts to take shape. All buildings respond in some way to the location and conditions of the land they are built on, but when you're building an eco-home, showing sensitivity to the land and its existing qualities and inhabitants becomes essential when considering your house's form, scope and materiality. The location, slope, contours, orientation and natural features like rock deposits, established trees and watercourses on your site will all suggest or even dictate what can be reasonably achieved in the placement and layout of your home. And the landscape isn't just restricted to the direct boundaries of the property — the wider physical and cultural context of your home may also guide your choices around how your house looks, so that it either blends into or plays off against the landscape in a pleasing way.

Once the construction work on your home is finished, the landscaping can begin in earnest (although if you can do some early planting in parts of the site that aren't in the line of access, this will give you a great head start), but don't leave the planning of the garden until the end of the building process. Planning the design of your garden at the same time as the design of your home will allow you to maximise the benefits of good landscaping, as well as the pleasure and well-being of all inhabitants of the site — human, fauna and flora.

It might seem that just because it involves plants, all landscaping is an innately 'green' activity. However, poor landscaping choices can disrupt local biodiversity, create water-run-off issues, leach chemicals into the soil and demand substantial ongoing maintenance to keep everything alive and looking good.

Taking a more natural approach to landscape design will minimise destructive interventions into the site while giving multiple paybacks: reduced maintenance (including using petrol or electricity for mowing lawns); reduced water demand by using rainwater and greywater; less need for fertilisers, pesticides and herbicides; greater biodiversity; improved soils; reduction of stormwater run-off; reduced electricity for air conditioning inside the home through using tree shade and waterscaping for passive cooling; and the pleasure and health benefits of eating fresh fruit, vegetables, herbs (and possibly even animal products) produced in your own backyard.

Natural landscape design doesn't mean that your garden must be wild and rambling, or exclusively native, or that it can't have any hard or artificial elements in it (like decks or pools). What it's about is working alongside nature, rather than against it. Award-winning landscape designer Xanthe White describes the garden as an enriching counterbalance to the home. 'I see the garden as a relief from architecture; it doesn't attempt to be architecture. Its softness and richness and wildness complements the control we have in our living spaces. I think natural gardens bring the clean lines of architecture to life and enrich them. It's a yin–yang thing, creating a tension between the two.'

Left: A dedicated space isn't necessary for a productive garden. At this Auckland home, Xanthe White was inspired by Moroccan courtyard gardens to interweave fruit trees, flowers, native ferns and creepers and herb plants to create a diverse garden with a 'human ecology'. Photo: Carme Aguayo

SOIL HEALTH

Developing the landscape around your home literally starts from the ground up. Whether you are developing an apartment or townhouse garden from scratch, transforming an existing site or repairing the disruption caused by building or renovation work, the place where it all begins is the soil, the foundation of all life in the landscape.

All around the world, soil is becoming dangerously depleted. Climate change, with its resulting droughts and flooding, is increasing localised flooding events in which the useful topsoil layer is washed away. The spread of cities, with their impermeable roads, footpaths and roof surfaces, increases this run-off problem. And as global demand for food increases, farmers rely more heavily on chemical fertilisers, pesticides and herbicides to keep their soils going — a quick-fix that shows decreasing returns over time. A UN-backed report released in September 2017 showed that 27 billion tonnes of fertile soil are being lost every year, and a third of all the world's soils are degraded. The UN has warned that at current rates of degradation, there may be only 60 years of harvests left.

The basis of health in the soil is the activity of the living networks within it. In the same way that we all have an internal microbiome of bacteria in our gut that helps us digest our food, soil has a complex microbial mix of bacteria, fungi, insects and worms that helps plants grow and uptake nutrients from the soil effectively. Key in this microbiome is the networks of mycorrhizal fungi that live on the root structures of plants. A good soil has a rich microbiome, high organic content, a high nutrient and mineral content, and the ability to let water move easily through it.

In many areas, this rich topsoil layer has been severely eroded — it may be barely a couple of centimetres thick — and on a post-construction site, it has likely been scraped back or compacted down to the point where it is difficult to get plants to thrive. So, the first job in restoring a healthy ecosystem is to get your soil back into good health. The exact way you go about this will differ depending on the type of soil you have on your property, but will likely involve a mix of composting, mulching, and feeding with nutrient feeds like worm castings and worm tea, living soil, biochar and soil separators like gypsum, lime and crushed rock. (Clean new plasterboard can be crushed into gypsym powder, so if you have waste plasterboard left over after building and clay soils that need help with water flow, using pulverised waste plasterboard is a great way of on-site recycling.) Soil also plays a significant role in carbon sequestration — so healing the soil also helps to help the atmosphere.

Sarah Smuts-Kennedy, an artist and biodynamic permaculture farmer who is in the process of restoring and transforming a formerly empty block of land in Mahurangi into a rich biodynamic garden, Maunga Kereru, suggests three simple strategies for nurturing agricultural soil during the restoration process:

Never stand on the soil you will use for growing produce in if you can avoid it. Standing on soil compacts it and crushes sprouting seedlings.

Never turn over soil once you have planted in it. Turning it breaks the mycorrhizal fungal webs that communicate across the soil and help transfer water and nutrients. Instead, use a fork to gently push it aside when planting.

Use 'chop and drop' composting. When your plant has finished fruiting or flowering, and is dying down (or if it is still going but needs pruning back), just chop it at surface level and lie it down to break down naturally. This creates a natural layer of mulch that will help prevent moisture evaporation. If the plant has gone to seed, eventually you will see it re-sprout in the soil without any additional effort on your part.

COMPOSTING

Composting is the act of breaking down organic matter so that it can be returned to the ground. It uses four elements — water, oxygen, nitrogen ('green' waste) and carbon ('brown' waste) to create an environment that's hospitable to microorganisms that can then biodegrade the raw materials into a nutrient-rich humus that can be used to fertilise plants and regenerate depleted soils.

There are numerous ways of composting: odourless kitchen-based systems that do most of the work within a contained box, 'tumbling' bins that rotate to help mix the compost and make it process faster, piles on flat land, in trenches or in wooden, plastic or metal bins or barrels, worm farms and hot compost boxes. If you live in an apartment or have a very small garden, a kitchen-based compost bin like a Bokashi system may be your best option. For larger gardens, outdoor piles, worm farms or box systems paired with a small bench-top or under-bench bin to collect kitchen scraps over two or three days (to save you having to ferry them out to the compost heap constantly) will generate more substantial amounts of compost.

Composting takes more than simply throwing food scraps on a heap of lawn clippings in your backyard. It's important to get the right mix of soft, rapidly degradable matter (like vegetable peelings, fruit scraps, coffee grounds, organic animal manure and green weeds) and woodier matter (like fern fronds, hedge trimmings, straw, shredded paper or cardboard), as well as water and heat to get the compost processing at an optimal level. When vermiculture (worm farming) is involved, certain food scraps can't be composted (such as onions and other alliums, meat, dairy and citrus fruits) because they disagree with the worms. Compost may need to be turned to keep it aerated, and exposed to rainfall to keep it moist.

WHAT IS LIVING SOIL?

Living soil is a form of biodynamic compost produced by blending organic cow dung, pulverised basalt and powdered eggshell into a smooth mix, spreading it in a covered pit and leaving it for three months. At the end, you have an incredibly rich, dark and nutrient-dense soil that contains more microorganisms in one teaspoonful than there are people on earth. It can be mixed with water to inoculate the roots of new plants by dipping them before planting.

WATER-SENSITIVE DESIGN

Water run-off is becoming a major issue to deal with in both urban and rural settings. Any hard surface — like roads, footpaths and roofs — that doesn't allow water to move through it and soak into the soil increases the volume of water that needs to move to a permeable area of ground or into a drain to be directed into streams, rivers or out to sea. As water moves over these hard surfaces, it collects sediment, chemicals, contaminants and rubbish.

As someone who has an urban stream running through their backyard, I see this with increasing frequency. After a big downpour, the quiet, clear little creek that's home to eels, fish and waterbirds can turn into a raging brown torrent 10 times its normal size, with the power to lift large tree stumps and send them bolting across properties at high speed. After every storm, people in the neighbourhood have to clear the stream banks of plastic water bottles, shopping bags, cigarette boxes and chip packets left behind as the water recedes. The topsoil is regularly stripped from the land, making it harder for new plants to establish and destabilising stream banks. In some areas, stormwater surges create sewage-system overflows, washing raw sewage full of pathogens and items flushed down the toilet directly onto beaches where people swim. At the most severe level, storms can cause flooding to homes and commercial areas.

PERMEABLE PAVING

Permeable paving allows water to pass through the paved surface, through a gravel layer, and into the soil below. As well as reducing overland storm-water run-off and pressure on drainage systems, water that moves through permeable paving into the soil is filtered of pollutants before it reaches the water table.

There is a wide range of approaches to permeable paving, both traditional and new. Cobblestones and Gobi block have been used for patios and

A Gobi-block driveway at a Herne Bay home is part of a wider natural garden design by Xanthe White that uses permeable surfaces to help manage water run-off naturally. Photo: Carme Aguayo

WHAT IS BIOCHAR?

Biochar is a charcoal soil enhancer that is produced by burning organic matter at high temperatures. In nature, biochar is produced during forest fires. Biochar absorbs and holds water and nutrients, decreasing the need for fertilisers, as well as absorbing and retaining chemicals and heavy metals, thereby purifying the soil and the water that runs through it. It also sequesters carbon in the soil for thousands of years. Biochar can be bought at gardening centres, or it's relatively easy to make your own at home (assuming you're not breaking any local bylaws around backyard burning). There are biochar interest groups throughout New Zealand that can advise on safe production of biochar.

driveways for decades, and provide hard-wearing continuous surfaces suitable for heavy use. Gravel and bark are also low-cost permeable surfaces for areas that are mainly walked on. Leaving gaps of a few centimetres between paving stones in a walking path also helps make the surface more permeable.

Porous concrete or paving stones have an open texture that allows water to pass through them and into the earth, although they are not as durable as pavers. Surfaces continuously paved with porous concrete or pavers need periodic cleaning to stop them becoming clogged with silt, mosses and decayed vegetation (such as leaf litter).

NATURAL CONTOURING

The way a site is shaped has a direct effect on how water flows across its surface. The usual approach taken in cities to water management is to try to make it disappear as much as possible by directing it into underground drains. But natural water flows can be a beautiful aspect of the landscape, enhancing biodiversity, greening the landscape, and bringing the musical sound of water flow to the environment. Studying the way water moves naturally across a site (and where it rests), and contouring to enhance it, can produce truly delightful results.

Xanthe White designed a swale to run down the site of a sloped urban site in Auckland. The area was originally designed to have a set of steel stairs above the run-off area, which would have needed retaining and piling. By working with the natural landscape, the design team created a beautiful, low-cost solution that is still a useable area during the drier months. Photos: David Straight (left), Natasha Iyer (right)

Water isn't a static element in the landscape. It recedes when the land and weather are dry, and expands when rainfall is high and the ground is saturated. It doesn't move in a straight line, but meanders, spreads and pools. A single garden might have low areas that are damp in the summer and very marshy in the winter, and high areas that drain quickly and dry out in the summer. Understanding how water flows across your property year-round can help you create a design to accommodate its changing nature.

A swale is an area of land that lies lower than the surrounding landscape and acts as a channel for water run-off. Over time, they tend to evolve naturally, but they can also be specifically designed to help manage water flows. They can be as simple as a ditch or boggy area lined with reeds and other wetland plants, or complex water paths that follow the swaying natural motion of water and include large rocks, trees and structures to allow crossings.

XERISCAPING

The inverse of run-off management, or the issue of dealing with too much water on a site, is water retention, or dealing with sites that receive only limited amounts of natural rainfall. Xeriscaping is a set of landscaping design principles that minimise the need for irrigation by using native plants that are adapted to the local climate, soil mulching, canopy layers and dense planting to prevent water evaporation, grouping plants with similar water needs together, and irrigation systems that reuse greywater, drip-feed instead of sprinkling, or use 'smart' control systems to measure soil moisture and dispense water only when needed.

ENCOURAGING BIODIVERSITY

New Zealand has around 1000 threatened species, according to the Ministry for the Environment, including all of our native frogs and bats, most of our native land snails and giant wetas, three-quarters of our native birds, more than a third of our native reptiles and freshwater fish, and hundreds of species of plants. Beneficial pollinators like bees and butterflies are also under threat from the varroa mite, wasp predation and honey-bee colony-collapse disorder, where otherwise healthy colonies of bees suddenly die with no apparent cause. Neonicotinoids, a category of agricultural pesticide, have increasingly been posited as a cause of these mysterious collapses.

Although backyard gardens are not complete habitats in themselves, they can play a vital role in preserving plant, insect, bird and animal biodiversity, particularly when they are linked with other neighbourhood gardens to create 'wildlife corridors', or continuous areas of shelter and food through which creatures and plants can spread.

And biodiversity benefits gardens and people, too. As any enthusiastic gardener knows, even the smallest gardens have microclimates that call for a diverse range of plantings. A plant that thrives in a sunny, rain-exposed spot can be near impossible to keep alive only metres away in an area that falls in the rain-shadow or sun-shadow of a large tree. Diversity is what gives an ecosystem its strength. A diverse ecosystem is also a lower-maintenance one, with different species performing different functions to keep the garden and its inhabitants in good health.

PLANTING NATIVES

Native plants are ideal for home gardens, not just because they preserve our unique biodiversity, but because they've adapted over millennia to be able to survive and thrive in New Zealand conditions, reducing the need for ongoing care and maintenance. Native plants attract native birds and insects — flaxes, kōwhai, karamū, karaka, mānuka and cabbage trees produce flowers and berries that attract tūī, bellbirds, silvereyes and native pollinators.

Of course, just because a plant is a native, it doesn't mean it will thrive in your particular garden — a coastal garden in Lyttelton will have quite different conditions to a rural garden in Northland. Exotic species also often fruit and flower at different times of year to natives, providing a year-round food supply for birds and insects. So there's plenty of room to interweave natives with exotics to create a mix that's aesthetically pleasing to you and well fitted to your garden.

LAWNS

A big, perfectly mown backyard, ideal for summer cricket games and sunbaking in lounge chairs while the barbecue sizzles, is an icon of Kiwi advertising, but it may be becoming a nostalgic relic of a time before suburban infill and climate change. Keeping a lawn lushly green and shorn short takes time, water and petrol (or electricity), as well as generating air pollution from the mowing and fertiliser and pesticide run-off.

Trading some or all of your lawn space for a productive garden of fruit trees, vegetable beds or flower gardens can reduce these impacts while adding direct benefits to your life — a garden that works hard, instead of creating hard work. Even if you can't give up your patch of flat lawn, you can make it work harder by sowing a diverse range of pasture species — clover, chicory, dandelion, comfrey — or a variety of hardy flowering ground-cover plants to create a 'tapestry lawn' that's attractive to bees and butterflies and will need less frequent mowing than a monoculture grass lawn.

AVOIDING CHEMICALS

Chemical insecticides are indiscriminate — they kill beneficial insect predators, like spiders, ladybirds and mantises, along with the pests they're designed to eliminate. Herbicides destroy life within the soil, but also permeate crops sown after the weeds have been defoliated — residues of glyphosate, the most widely used herbicide, have been found so pervasively in food products, topsoils and even human urine samples that the European Union has heavily debated a Europe-wide ban. And fertilisers can run off into groundwater and streams, causing algal blooms that smother water life and are a hazard to animals and children.

A limited number of organic-approved natural pest-control solutions exist, such as neem oil, citrus oils, castile soap mixes and diatomaceous earth. However, although these are gentler on the environment, they're still pretty indiscriminate on insects, so improving your soil health through regenerating topsoil is perhaps the best thing you can do to eliminate chemical use. Plants that grow in rich soil do more than fruit and flower well — they're naturally more pest-resistant, too.

Hand-weeding is still the most effective method of weed control for backyard gardeners, although two new commercial weed-control methods that have both recently arrived in New Zealand — one using saturated steam and boiling water, and one an electrocution control method known as 'hot grass' technology — show promise.

WILDLIFE HABITATS

As well as planting year-round food sources for native birds and animals, you can also create specific shelters within your garden for creatures that sometimes struggle to find suitable homes for themselves. Bumblebee queens spend their winters opportunistically sheltering in abandoned burrows, and under leaf litter. They like to establish hives in early spring in crevices or holes in trees. Nesting boxes are easy to build, or can be bought online.

Native geckos and skinks like hiding places that they can dart to easily — hollow logs or large pieces of wood slightly raised above the ground. Forest and Bird suggests building a 'lizard hotel' by layering squares of corrugated roofing iron or tiles with small pebbles between the layers to create slim gaps that the lizards can slip into. Put the hotel in a sunny place with nearby tussocks or grasses, flat rocks for sunbathing on, and a shallow bowl of drinking water.

Backyard beekeeping is also growing in popularity in urban areas. Some commercial beekeepers offer a hire service where you can rent a hive for your backyard, which the beekeeper will check and maintain, and deliver the honey to you after harvest time.

PRODUCTIVE GARDENS

Eating more plants is widely recognised as being good for personal health, as well as reducing load on the planet. In the book *Drawdown*, Paul Hawken ranks a plant-based diet as the fourth most effective solution to climate change. New Zealand's climate emissions are one of the highest per capita in the world, in substantial part due to the emissions created by dairy, beef and sheep farming. Transitioning to a more plant-based diet addresses a distinctly local environmental concern — and one of the best ways to get more vegetables and fruit into your diet is to grow them yourself.

There's no shortage of advice available on getting started in food gardening, but one system that's designed to provide maximum results for the least ongoing effort is permaculture. Conceived by Australian agriculturalists Bill Mollison and David Holmgren in the 1960s, it emphasises creating sustainable and efficient systems that minimise waste and effort by carefully observing your local ecosystem and designing your garden to cluster plants that work well together in 'guilds', and take advantage of naturally productive zones like edges. It's a design system that can be applied to architecture as well as agriculture.

BUILT-IN LANDSCAPING

DECKS AND PLAY EQUIPMENT

Outdoor decks need to be hard-wearing because they take a lashing from sun, wind and rain, and can be exposed to insect damage. Using treated soft timbers like pine for decking (or children's play equipment) runs the risk of exposing bare skin to harsh chemicals when you walk, sit or lie on your deck, as well as chemical run-off into the soil. Accetylated wood (such as Accoya), thermally modified wood (such as Abodo) or naturally durable timbers like cedar or macrocarpa are better choices without the same risk of chemical exposure. Staining using exterior stains based on natural oils and resins will further improve the durability of the timber.

Imported hardwoods like kwila (or merbau), Fijian mahogany, jarrah, vitex and garaba (Brazilian oak) are naturally durable, but are often sourced (sometimes illegally) from rapidly diminishing old-growth tropical rainforest. If you're considering this kind of wood, it's essential that it has a highly regarded international certification like FSC or PEFC. Local certifications may not have the same sustainability standards. It's also worth considering the carbon kilometres associated with importing these woods.

Composite decking, or wood–plastic composites (WPCs), are becoming more popular as a decking material. Using recycled wood chip, rice husk and plastic formed into long boards, they're often pitched as a sustainable alternative to real wood because they 'don't involve cutting down trees'. This claim only has merit when compared to unsustainably harvested trees — there's no environmental issue with cutting down trees that will be replanted. In addition, composite decking products contain glues, may be chemically treated, and don't have end-of-life recyclability or proven biodegradability. As they're a relatively recent entry to the local market, their long-term durability hasn't been established in New Zealand's climate, although they've had good results overseas. Look for WPCs that are 100 per cent recycled material, and certified by an external agency, such as Greenguard.

This natural pool, installed by Natural Pools NZ in Kaiteriteri, has two planted 'wings' that keep the water clear without any need for chemical or salt treatment. Photo: Natural Pools NZ

GREEN WALLS

Green walls (also known as living walls or vertical gardens) are a great option for outdoor areas that are limited in size. As well as helping absorb storm-water run-off, shading walls and reducing heat and noise, they can be used as productive gardens.

Exterior green walls are a much more manageable choice than interior living walls, which require a lot of ongoing maintenance to keep them in good condition. It's important to choose your plants carefully for the kind of sun and wind exposure that they'll get, and if they don't get enough natural rainfall, install irrigation systems that are gravity-fed rather than relying on electric pumps, and only build to a height that makes maintenance simple. Making sure there's room for air to move between the plants and the wall surface will reduce the risk of moisture and mould problems.

NATURAL POOLS

Natural or chemical-free pools use plants as a filtration system to create a home pool experience that's like swimming in a freshwater lake or pond. They've been available in New Zealand for close to a decade, and are slowly becoming more popular, with several companies offering them as an option.

One of the big misconceptions around natural pools is that they're more expensive than standard pools, when they actually cost about the same to install, and may have reduced ongoing maintenance costs. Existing pools can even be converted to natural pools.

Natural pools are separated into two areas — a deep, clear one for swimming, and a shallow stone-lined one that can be planted densely with reeds and other water plants that will grow up to (and even beyond) the surface of the water. The plants filter the water by absorbing nitrates and ammonium, and release oxygen. When the planting is extended beyond the edge of the pool, it can blend beautifully into the landscape. A pump system (which can be run off a solar panel) recirculates the water through the pool. Small fish can even be added to the pool to keep insects under control.

A METAPHOR FOR LIFE

More than anything, gardening is life in microcosm: dreaming and planning; careful nurture; the delightful blossoming of new life; despair and fury when unexpected plagues and disasters devastate a project you've put time and energy into. The abundance teaches us about nature's generosity; the failures teach us about patience and the interconnectedness of our ecosystem, and give us a deeper understanding of how the things we design can work to either support or undermine the place where we live.

Buildings allow us the temporary illusion of dominance over nature, that we can construct a shelter against the elements. But as time goes on, it becomes clear that even the most solidly constructed eco home must be well designed and carefully maintained to resist the relentless forces of nature over time. Spending time in close interaction with the living world helps us understand that there are limits to what we can achieve, and the benefits of living and working in harmony with nature, rather than in competition with it.

WEBSITES

www.beaconpathway.co.nz
Excellent sustainable building and living advice, research and reports and services.

www.branz.co.nz
Building advice and in-depth reports of all types from the Building Research Association of New Zealand. Also see www.level.org.nz for specific sustainable building advice.

www.ecodesignadvisor.org.nz
Advice and guides for making your home healthier, warmer and more sustainable.

www.energywise.govt.nz
Advice and online tools to help you to reduce your energy use.

www.greenbuildingadvisor.com
International sustainable building website — great general advice and information on key principles like passive design, but some information may not apply to the New Zealand context.

www.nzgbc.org.nz
Home of the New Zealand Green Building Council, with advice on Homestar and professional recommendations.

www.smarterhomes.org.nz
The Ministry of Business, Innovation and Enterprise's guide to improving the performance and sustainability of your home.

www.sustaintrust.org.nz
Wellington-based sustainable homes advice and online shop.

www.recycle.co.nz
Information on how and what to recycle within New Zealand.

The author would like to acknowledge the wonderful research work of BRANZ, Beacon Pathway, the New Zealand Eco Design Advisor service, the New Zealand Green Building Council and the many architects, building and ecology experts and organisations who contributed their time and knowledge during the research process for this book. Any errors are entirely my own.

FURTHER READING

Alexander, Christopher, Ishikawa, Sarah and Silverstein, Murray, *A Pattern Language*, Oxford University Press, 1977

Alexander, Christopher, *The Timeless Way of Building*, Oxford University Press, 1979

Blackmore, Andrea and Easton, Lois, *Making Your Home Smart: A Homeowner Manual*, Beacon Pathway

Bolchover, Joshua, *Vitamin Green*, Phaidon Press, 2012

Conran, Terence, *The Eco House Book*, Octopus Publishing, 2012

Gunders, Dana, *The Waste-Free Kitchen Handbook*, Dana Gunders, 2015

Hawken, Paul, *Drawdown*, Penguin Books, 2017

Hosey, Lance, *The Shape of Green: Aesthetics, Ecology and Design*, Island Press, 2012

Leopold, Aldo, *A Sand County Almanac*, Oxford University Press, 1949

McDonough, William and Braungart, Michael *Cradle to Cradle: Remaking the Way We Make Things*, Farrar, Straus and Giroux, 2002

McLennan, Jason F. *The Philosophy of Sustainable Design: The Future of Architecture*, Ecotone Publishing, 2004

Naisbitt, John, High Tech, High Touch: Technology and Our Accelerated Search for Meaning, Broadway Books, 1999

Roberts, Lin, et al., 'The Nature of Wellbeing', Department of Conservation, 2015

Sachs, Jeffrey, D. *The Age of Sustainable Development*, Columbia University Press, 2015

Shaw, Peter, *A History of New Zealand Architecture*, Hodder Moa Beckett, 2003

White, Xanthe, *The Good Dirt*, Penguin Random House, 2016

Wilson, Edward O., *Biophilia*, Harvard University Press, 1984

INDEX